FIXOLIK®

A TO Z

One Stop Solution of Waterproofing

FIXOLIK® A TO Z

One Stop Solution of Waterproofing

Authored By,

Maulik Choksi

WINGS PUBLICATION INTERNATIONAL

WE GIVE WINGS TO YOUR DREAM!

Disclaimer

WINGS
PUBLICATION
INTERNATIONAL
WE GIVE WINGS TO YOUR DREAM!

Registered Office- 907-Sneh Nagar, Sapna Sangeeta Road,
Agrasen Square, Indore – 452001 (M.P.), India

Website: http://www.wingspublication.com

Email: mybook@wingspublication.com

First Published by WINGS PUBLICATION 2024

Number of Copy : 500 Nos

Title : FIXOLIK A TO Z: One Stop Solution of Waterproofing

Price : Rs. 1649 | $ 25

ISBN : 978-93-6006-793-9

LIMITS OF LIABILITY/DISCLAIMER OF WARRANTY

Dedication

My Loving Parents,

I acknowledge my deep gratitude for your enormous love, affection & selfless efforts in the pages of this book called "Fixolik - A to Z – One stop solution of waterproofing", you are my inspiration for all the time. Through this book I once again express my gratitude for everything that you have done for me.

With all my love & affection,

Maulik Choksi.

Acknowledgement

With the great support and enthusiasm shown by all my near & dear colleagues, family members etc., I was inspired beyond my expectation to write this book on my profession.

To my beloved wife, Bhumika (Sweetu), you have been my constant supporter, my confidence and my wings. Your patience & understanding and countless discussion cannot remain unnoticed and cannot be forgetted forever. Without your guidance & motivation, it will not have become possible to write this book.

Hetal & Milap, my dearest sister & brother, I cannot forget your support & all the best wishes for me, forever.

Dhyan & Ahan my loving children, Your Presence of mind at all the time & giving inputs to me at the very small age has really inspired me. I give my 100% to both of you for your selfless support in publishing my book.

Mr. Dipak M Shah, not only my father in law but a true friend, philosopher & guide for me. Your selfless sacrifice made my journey of writing this book, very easy and gave me enough confidence.

Mr. Kailash Sir & Deepak Sir, a mentor of his own outstanding class, has been true guide and has doubled my confidence and pushed me correctly to write this book.

Mr. Sneh Desai, your live & unforgettable sessions have given me marvellous support in writing this book. I deeply acknowledge your constant effort in promoting my thoughts throughout my business career. I hereby take this opportunity to thank you for your selfless prop up in preparing foundation of my career.

To all my friends & colleagues, past & present from whom I learn a lot. Their experience has helped me in framing my career. I am extremely grateful for their support in all respect.

To everyone, who has been a part of my journey whether as family, mentor, Team FIXOLIK (colleague), etc. has left unforgettable impression in writing this book.

Thank you all once again,

With sincere gratitude & regards,

Maulik Choksi

Preface

This book is basically a road map for all Civil Engineers & individuals working in construction and as specially in a waterproofing industry.

This book will be very useful for whom who are starting their career in waterproofing profession. Actually this is a very big industry not only in India, but also at world level. If your house or commercial property is leak free than you will get inner satisfaction of constructing this house or office property.

Through this book my intention is to inspire the readers, young professionals & others to achieve the goal of leak proof India & leak proof world at a large.

After reading this book, I am sure that the person will get all the ideas of how to do waterproofing work even in the most difficult situation. This will give overall satisfaction & doing proper & justified book writing work.

I submit that, I have done my best in giving all the material information either technical or non technical for better understanding of most difficult waterproofing subject.

I have an experience of around 20 years in waterproofing work. That is why I have given useful ideas & techniques (of waterproofing at various places) learned during my career. I have made an honest try for giving as much knowledge as possible.

Through this book, I am giving something back to the engineering community that has helped me to come at this stage of expertise in waterproofing. I want to motivate & encourage other Civil engineers and waterproofing contractors to do their work in a much more easy manner.

Ultimately, my goal is leak proof India & leak proof world.

With Warm Regards,

Maulik Choksi.

Foreword

For readers, this book is a perfect book on waterproofing. Mr. Maulik choksi's life (about 25 years) experience reflects in it. If anybody interested in doing the business of waterproofing with the support of this book, he can singly do his best.

Students & Professionals will also have enough guidance for transformation of their life.This book opens a new door for water proofing industry. One can become master of it by going through properly. You can face any challenge whether technical / professional or any other, after deep study of this book.

Now get ready for success!!

Sneh Desai,

India's Leading Life Coach & Business Coach,

International Best Selling Author.

About The Author

Mr. Maulik Choksi is a founder & promoter of the well known firm shreeji construction.After obtaining his degree in B.E. (Civil Engineering) from M.S.University, Vadodara, he commenced his career in construction industry. Since last more than 25 years, he is associated with **Waterproofing & Structural Repairing work.**

He is fully dedicated to elevate the industry by imparting knowledge & expertise. His focus is also on imparting training & mentoring. Uptill now he has trained 2000 new entrants with a particular emphasis on advancing the field of waterproofing & structural repairing.

He is sharing his experience through Social Medias. His **You Tube channel** is well known on the name of "**FIXOLIK**". He has produced 95+ videos, setting a record as the highest in India. His vision is "Leak proof & Healthy India Mission" (LP & HIM) by making 10,000 buildings leak proof & structurally sound by 2030.

Further beyond business venture, his **You Tube channel, "Maulik Choksi"** is also very famous. His title "Meaning full moments with Maulik (MMWM)" is focusing on Time Management, Goal Setting, Meditation, Yogasana & Pranayam, Parenting, Health show & many more.

Further, he is offering his services in

1) Water proofing consultancy (PAN India – For pioneer concept, Complete PMC of Waterproofing from foundation to finishing)

2) Structural Strengthening /Repairing work consultancy (PAN India)

3) Online & In person waterproofing Training start up program / Module.

He is also associated with

(1) The Gujarat Institute of civil engineers & Architects (GICEA) FLM No.4135

(2) The Water Proofers Welfare Association (He is co founder & core committee Member),

(3) The Water Proffers Association of India (WAI),

(4) 21 Gaam Ichhaben ni vadi (President), Ahmedabad

(5) Member of SuperPreneur Club of India

(6) Member of Real Estate Club of India

(7) Core Committee member of "Positive Jindagi" &

(8) Core Committee member of "Team On - Business Networking Group"

Mr. Maulik choksi with his brand name "**Fixolik - Fix the leak with Maulik**", Provides the following services as a consultant.

1) Project Estimate
2) Specification
3) Drawing
4) Technical Training / Presentation
5) Waterproofing Experienced Expert Agencies/ Contractors
6) Mock on site
7) Site Work Tracking
8) Certification
9) Creating WhatsApp group for smooth day to day operation.
10) Service Provider for Technical Staff (Optional)

His Clients are,

1) Ganesh Housing Corporation Ltd
2) Gala Group

3) Arvind Smart Spaces

4) Vishwanath Builders

5) Apple woods

6) Sandesh

7) Savvy

8) RBL

9) Infra Projects Pvt Ltd.

10) Meera Group

11) Vastu Shilp consultants

12) Artech Engineering Solutions

13) Pandya & Associates

14) Eyecon

15) Western construction Group

16) Shilp Building Excellence

17) Ethios Enviro solutions (P) Ltd.

18) College Design

19) AIA Engineering Ltd.

20) Adroit Techno Engineering Solutions.

21) TATA motors

22) Chiripal Industries Ltd.

23) TMVT Industries Pvt. Ltd.

24) MPPL

25) Shree Narayana Cultural Mission

26) Mount Carmel School

27) Vruksha Landscapes

28) Amneal pharmaceuticals

29) Synergy Neonatal & Pediatric Hospital

30) SK Group

31) Shree Raj Saubhag Satsang Mandal

32) Ardex Endura

And many more

Now, with the sole intention of giving something back to the industry, this book is published in the larger interest of engineering community.

Journey of FIXOLIK

Today will see how Maulik Choksi has prepared a base for waterproofing through his leak proof healthy India Mission.

In fact, he has added his 19+ years experience in this book named "FIXOLIK A TO Z One Stop Solution of Waterproofing". This book will prove very useful right from students to each and every body engaged in waterproofing business or wants to start his new business in the field of waterproofing.

There is a story behind this book, Mr. Maulik Choksi was born in a rich family, i.e. with golden spoon in his mouth, but he lost everything at the age of 19. The only thing left with him is to complete his education and take the responsibility of family.

He selected a civil engineering as his career. His dream was to become a Builder, but because of shortage of funds, he did not become Builder. Now he thought, what is to be done to give the best reward to the society and also to earn for his family.

Mr. Maulik Choksi has 2 skills i.e. professional and spiritual. He is also a very good motivator & has to impart his knowledge. He started to prepare videos on motivation & on his waterproofing business.

During Covid19, he helped a lot of people through his motivational videos. He also helped to the people who were very much depressed about the running of their business. He prepared around 90+ motivational & professional videos & uploaded all 90+ videos on social media, i.e. YouTube, but he was unaware about the future & one day his all 90+ videos were hacked and all videos are now unreachable. You can think how many hurdles he had faced at that time but mentally he was very strong & without changing his path, he decided to continue his work that what he has decided to do.

Behind this thought & behind his strong will power, the person responsible is his coach & mentor who increased his knowledge & power to sustain. That is why today, Mr. Maulik Choksi wants to express his gratitude towards all his coaches & mentors. The whole names are as under.

1) Shri Shri Ravishankar Maharaj of art of Living

2) Life & Business Coach Shri Sneh Desai

3) Communication & Success Digital Marketing Coach Mr. Nakshatra Meuva

4) Coach Mr. Sandeep Gupta

5) Coach Mr. Dev Gadhvi

6 Motivational Speaker Coach Mr. B.S.R

7) Owner & Founder of "Vishala" Mr. Surendra Patel

8) Humble man Mr. Sonu Sood

9) Great Actor leading & businessman Mr. Abhishek Bachchan

10) Self development & Mind Power Trainer Mr. Jitendra Adhia

After taking training from all these great personalities, Mr. Maulik Choksi started his PAN India Consultancy of waterproofing. This is today's necessity of India.So friends, Stay with "Fixolik", Stay with Maulik Choksi, "Fixolik - Fix the leak with Maulik Choksi".

Please scan the QR Code for further information

"Life to me is a journey - you never know what may be your next destination."

- David Russell

Dr. A.P.J Abdul Kalam Inspiration Award 2024

Category : Most Inspiring Leader of The Year 2024 (Waterproofing Industry) With Ms. Hema Malini (Bollywood Actress & Member of Parliament) & Mr. Rameshchandra Ratna (Advocate Supreme Court & Chairman, P.S.C., Ministry of Railways, Railway Board, Government of India

Testimonials

Shreeji Construction & Consultants rendered their services as per project requirements and schedule with our complete satisfaction. Their insights and guidance were invaluable. Team 'FIXOLIK' at the waterproofing consultancy demonstrated professionalism and proficiency, delivering exceptional results that have made a significant difference for our property. It has been a distinct pleasure to work with Mr. Maulik Choksi and team FIXOLIK.

– Ganesh Housing Corporation Limited, Ahmedabad

With my 48 years of experiences, I can say that this book contains all the documentation/information on waterproofing and it is to be preserved for help at any time. Future systems of waterproofing are applied in the present. Mr Maulik Choksi is an expert in digitalisation.

- Mr Kalpesh Shah, Director, Vipul Waterproofers Pvt Ltd.

Maulik is a Gujarati word, and if you translate it in English, it means 'Original.' Maulik Choksi is also very original and pure like his name. He is a person who is always having high vibrations, creating a positive environment around him. He believes in delivering more than what he promises for. I know him for many years and I can say that Maulik Choksi is a person to get associated with. He has mastered the art of waterproofing. This is what you are going to experience from his book. His book is going to be proved as a landmark in the waterproofing industry. All the best.

- Manoj Meuva, Naxatra Vasstu World, Vastu Expert, Numerology Expert, Swar Vigyan Expert

Mr Maulik Choksi is an avid learner. He has a nature of looking at things with a very open mindset. He is an individual who embraces change very swiftly on the go. He is very helpful, caretaking and a kind hearted person. As a business coach, I have explored many business models of manufacturing, trading or service industry, but during my physical interaction and after reading the book authored by Mr Maulik Choksi, I can say that he has not left a single stone untouched about waterproofing in this book, and it will really be a privilege and pleasure for all the readers who are directly and indirectly associated with waterproofing industry.

- Mr Sandip Gajjar, Director & CEO, PDV Education Pvt Ltd.

This book is a treasure trove of knowledge, filled with practical insights and comprehensive information on all types of waterproofing work. Whether you are just starting your career in this field or are a seasoned professional, the wisdom shared in these pages is invaluable.

– Dr Kailash Pinjani, International Author Coach

Mr Maulik Choksi is a knowledgeable professional in the field of waterproofing and structural repairs since 2005. His expertise & passion for the industry are truly commendable. With my more than 33 years of waterproofing industries experiences, I recommend that his book is a comprehensive guide for anybody looking to understand and excel in the field of waterproofing & structural repairs. I wish him all the best with the launch of his book & look forward to see a positive impact in the industry.

– Mr. A J Charles, Director, Perma Constructions Aids Pvt Ltd.

What sets this book apart is not just the technical expertise but also the inspiring personal journey of Maulik Choksi. From a determined student to a veteran professional, Maulik's story is a testament to growth. His vision for LP&HIM highlights his commitment to the profession and to society.

– Dr Deepak Parbat, Director, Wings Publication International

Mr Maulik Choksi's dedication and expertise in the field of waterproofing are commendable. His innovative approach and commitment to quality have left a lasting impression on me. His contribution to the industry is invaluable. The book encapsulates practical wisdom and industry insights, making it a must read for anyone passionate about waterproofing solutions, this I can definitely say that by my more than 30 years of industry experiences.

- M K Dhawan, CEO, Dhawan Associates

To become a great teacher, you have to become a great learner. If I have to summarise Maulik Ji's personality then I can surely say that he is someone who keeps on updating himself; not only that, he makes sure that the people around him are also constantly learning and upgrading. Maulik Choksi, someone who always does things in his own way and creates a pathway for the future generations, just like he has been a pioneer in the waterproofing industry. Anyone willing to know more about the inside-out of waterproofing then pick this book up without a doubt.

- Mr Nakshatra Meuva Communication & Relationship Coach, Sneh Academic Services Pvt Ltd, Ahmedabad

Mr Maulik Choksi is a renowned waterproofing consultant in Ahmedabad & Gujarat with fair practice & with new concept. He is doing waterproofing since more than 20 years. He has given lots of new ideas on this subject. He is very soft spoken. The book launched by Mr Maulik Choksi is worth to study even for waterproofing experts. He has incorporated all kinds of episodes of effective waterproofing.

- Mr. Kamlesh Shah worked as Br. Manager to AGM and Technical Head, Choksey, Cico, Pidilite, Kamdhenu, STP, MAPEI & K2 (More than 25 years of waterproofing industry experiences).

I know Mr Maulik Choksi since more than 15 years. He has good technical know-how about construction, chemicals, products, and putting his best efforts to serve the industry by spreading awareness through his You Tube knowledge-sharing sessions on various topics. My best wishes to Maulik Choksi. This book will be a path finder to all enthusiasts who want to learn about construction chemicals.

- Mr Bhairav Prajapati, Product Specialist, Pidilite Industries Ltd. (24 years of waterproofing industry experiences)

Maulik Choksi's latest book is nothing short of a masterpiece for anyone involved in the construction industry, particularly those specializing in waterproofing. As a qualified B.E. Civil Engineer, Maulik dives deep into the technical nuances of waterproofing in a way that is both captivating and immensely educational.

- Monika Singh, CEO, Wings Publication International

Mr Maulik Choksi is a well-experienced person not only in giving specifications, but also in application areas, too. He is a very good trainer as well as instructor on the subject of waterproofing. His devotion is wholehearted and provides any kind of solution related to the waterproofing subject. I wish him all the success for his consultation in waterproofing and also his new introduction of book on waterproofing, which will help the entire construction industry for making leak-proof buildings across the nation. Salute to Maulik Bhai. I insist that everyone must read this book by my 23 years of industry experiences on waterproofing.

- Mr. Yogesh Kulkarni, Ex. DGM Sales, Zonal Head - Central & North Region, Ardex Endura India Pvt Ltd.

This book is an excellent and first-of-its-kind in waterproofing industry. It is a masterpiece and sharing whole knowledge of waterproofing in a single book. Mr Maulik Choksi is a very knowledgeable and ethical person. His approach is value based, and he is also very socially active.

- Mr. Alankar Wakode, Cluster Manager & Gujarat Head, Myk Arment Pvt Ltd.

Contents

Chapter

1

Waterproofing Consultant: - Need of Today

What is Waterproofing?

Waterproofing is an important process/coating system that prevents water from entering the structure/mother surface of the construction.

1.1 Role of Waterproofing Consultant

- A well-built building structure is often not paid attention to, and it does not show any failure of minor or major deficiencies, but the results are often very harmful.
- Waterproofing Consultants play an Important Role & make the project service long life & save money.
- Extra assurance Quality Check by Consultants.

❖ In the Construction Industry, how can consultants play a role?

1. Preconstruction Support
2. Post Construction Forensics

1. Preconstruction Support:

During the Excavation Work Starting/Project launching time, the consultant must be required from the foundation to finish the work of the entire project. Builders, developers, architects, interior designers, industrialists, waterproofing distributors, realty companies, hotel industries, archaeological departments, Project Management Consultants, Trunky Contractor, MEP Contractor (Plumbing), Structural Engineer, Government projects, colleges & any other structures in infrastructure will hire consultants for waterproofing. After that, the consultant will meet the architect & interior and discuss the design of the entire project work and thereafter decide which type of materials are to be used for which area so that they can complete proper work without any deviation. Therefore, the project is completed on time, which saves your time, money & energy, and finally, you will get the best service life for your buildings/projects where there are minimal chances of leakages.

2. **Post-Construction Forensics:**

In Post Constructions, where structures/building has been ready or completed, but there are defects/failures in structure. Thereafter they might be needed to advise of waterproofing consultant for finding a route cause of defects/ failures in structure, and suggestion for solutions/remedies for entire area.

Waterproofing consultants work commence with moisture detection. Therefore, a consultant will find the root cause of the main point of leakage with his expertise and his check on-site, thereafter finalising the design/system for the same.

Electronic Leak Detection System (Thermal Detection Camera) have to be used for identify a leakage area by consultant, thereafter his designed for solutions of the entire area. Consultant will give you a quality assurance for the waterproofing system.

If you are hiring a consultant, they will manage overall waterproofing work with their expertise, and you will be free from any kind of trouble of waterproofing. Waterproofing consultants use advanced tools/equipment, as it might not overlook any water damage. Consultants will work in a systematic manner (i.e. Plan layout, overall steps, Quality checkpoint, etc.). Therefore, it might give you proper work and give your money worthiness. Therefore, it will save your time, money & energy. In addition, also they might be creating a WIN: WIN: WIN condition for the Owner, Consultant & Waterproofing Contractor/Agency.

Please scan the QR Code for further information:

"The road to success and the road to failure are almost exactly the same."

- Colin R. Davis.

1.2 Waterproofing is necessary in construction work

In Building Construction/Structures, Waterproofing must be required as construction time masonry & concrete were not water tight, so it is required to add some additives and then it gives you one level quality for water tightness.

Waterproofing is a barrier between Water & building structures to prevent moisture content. Conventional waterproofing systems use sheet/liquid membranes for waterproofing (i.e., bitumen type, coal tar type, PVC, EPDM, cementitious, acrylic, etc.). It may depend on the selection of the waterproofing area.

Why is Waterproofing required in the present scenario,

1. **Global Warming**: "Climate change" refers to the increasing changes in the measures of climate over a long period of time – including precipitation, temperature and wind patterns. It may affect/damage building's' surfaces/structures.
2. **Excellence in workmanship**: Awareness of workmanship in labour or less expertise in work or Speed of projects affects the workmanship of labour.

Owner/Project Engineer may invest a lot of money that is not necessary & it directly affects investment money that is actually required for waterproofing but cannot be used.

Why Waterproofing Consultancy has to be crucial?

In Building Structure, i.e. Expansion Joint, Construction Joint, Cold Joints are mainly points of water ingress and it may create a more chances to damages in structures. Therefore, it might be required to perfect solutions for damages in building/structures. A skilled agency is essential for this kind of work, but an ordinary agency can achieve excellent work with the knowledge of an expert/consultant.

In Podium Slab Waterproofing, the Basement Area/Parking Area/Driveway Area has vulnerable points that let the water into structures/components of buildings. Fountain areas may be points of water ingress into the structures. If the Basement area is not treated for waterproofing, it may affect the structures of buildings & service life of buildings.

Raft foundation is the foundation of the building, therefore waterproofing is most crucial, as there are chances to Raft uplift in monsoon season (Water Level up down in Season) as waterproofing is not done properly. So, an expert can be hired to do the best work along with cost control.

Retaining Wall Area/Diaphragm Wall Area must be done with proper waterproofing as it might create issues of dampness, capillary action or rising bar dampness & it may affects the service life of buildings. So, an expert can be hired to do the best work along with cost control.

Now, with the RERA term, Builders will provide a 5 year warranty to clients. So, it might require consultant's suggestions towards best waterproofing systems as well as site supervision & execution of entire works with his best expertise. Thereafter, the consultant will insist a 5 to 10-year warranty as per area & quantum of work after discussion with the Contractor/Agency, also provide triparty warranty agreement with Waterproofing Manufacture Company for last-long relationship in this Waterproofing industry. Therefore, Buildings/Structures are safe from any internal damages & also maintain a better service life of it.

Please scan the QR Code for further information:

"Success usually comes to those who are too busy looking for it."

- Henry David Thoreau

1.3 Benefits of Waterproofing Consultant

❖ **Benefits of Consultants:**

1. Helps you to avoid costly mistakes & save money

Waterproofing consultant will provide better products for waterproofing work at its expertise and control a value of overall projects.

2. Help you in utilising your budget efficiently.

Waterproofing consultants will give a better suggestion than any other person as they have a lot of experience in the waterproofing field, so he may be designed properly for which products to be used for which area and also have knowledge about where to invest the money. In addition, he has knowledge of how to complete overall work with his best expertise.

3. Save your Money & Energy

Consultants have a wide experience/expertise in waterproofing field, so they can complete a waterproofing work within timeline of project and also work on planning accordingly. Consultant will give you quality work where waterproofing is not required for rework in working area. So, it can save your money, time & energy.

4. Build a strong bridge between the client & agency.

Waterproofing Agency/Contractor have different mindsets

and different strategies for working in this field. If you have hired a consultant, they have built up a strong bridge between client & agency as they have wide experiences of how to deal with different multiple agencies in rate finalisation, methodology for work.

5. **Have a wide availability of resources.**

Many times, construction work is to slow down due to lack of resources/agencies, etc. If you will work with consultancy, consultants can provide many experienced contractors, suppliers, wide range of availability, etc. Therefore, the project is complete within stipulated period. Thus, consultants can save all resources for entire work.

6. **Helps you give the "Wow" Factor**

Consultant will work with proper checklist & guidelines, so it may help you to complete work within timeline & gives a best quality of entire work, thus it may reduce to chances of leakages in structures/buildings. Thus, it may provide a "wow" factor.

7. **Add a great deal to the aesthetics of your products/ property to raise its market value.**

Waterproofing is a concealed part of your building/structures, so it may be required to provide the best waterproofing system. Thus, you can hire a consultant for overall project waterproofing work. It may help you to provide the aesthetics

& security of buildings/structures. Therefore, it may provide a long service life to buildings/structures, which may result in increase of property market value.

8. **Consultant's trained eye tells you if there is something wrong or right with a space.**

Consultant has a wide experience/expertise of this field. If you will work with unwanted failure system then it may be check & guide how to resolve matter and provides a best solution for it.

Therefore, a Waterproofing consultant is necessary for waterproofing work in every buildings/infrastructures.

Please scan the QR Code for further information:

"We cannot solve problems with the kind of thinking we employed when we came up with them."

- Albert Einstein

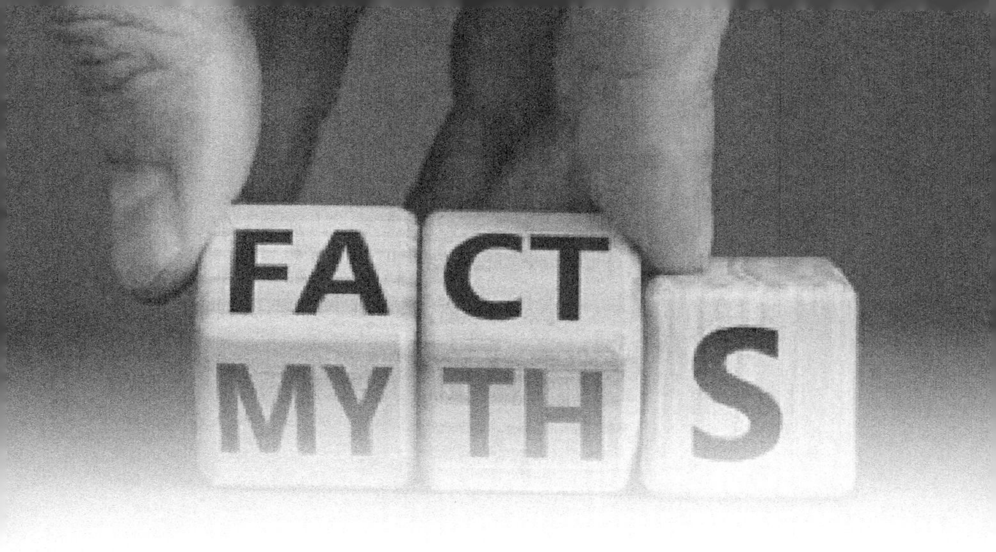

Chapter

2

MYTHS – Waterproofing

Here are 25 common Myths for waterproofing:

1. **Sealing cracks is enough**: While sealing cracks is important, it may not be sufficient to fully waterproof a structure. A comprehensive waterproofing system is often needed.

2. **Waterproofing is only for buildings**: Waterproofing is also important for other structures like bridges, tunnels, canals, archaeological monuments, Industry, airports, Docks & Harbours, Cement plants, Steel Plants, parking garages & many more areas.

3. **Waterproofing is only necessary for basements**: While basements are commonly waterproofed, other areas like roofs, bathrooms, and foundations also require waterproofing.

4. **Waterproofing is only for new construction**: Waterproofing can be done on existing structures as well to prevent water damage and leaks.

5. **Waterproofing is only necessary in wet climates**: Even in drier climates, waterproofing is important to prevent water damage from occasional rain or plumbing leaks.

6. **Waterproofing is expensive**: While waterproofing may require an initial investment, it can save money in the long run by preventing costly water damage repairs. In long run waterproofing is an investment & not expenses.

7. **Waterproofing lasts forever**: Waterproofing materials can deteriorate over time and may need to be replaced or repaired periodically, almost all waterproofing application also have certain years of sustainability not last forever.

8. **Waterproofing products are all the same**: There are many different types of waterproofing products available, each suited to different applications and environments for different areas.

9. **Waterproofing is only for below-ground structures**: Waterproofing is important for both below-ground and above-ground structures to prevent water damage during service life.

10. **Waterproofing causes mould**: While improperly installed waterproofing can trap moisture and lead to mould growth, properly installed waterproofing can actually prevent mould by keeping moisture out.

11. **Waterproofing all membranes are the same**: Waterproofing membranes vary in material, thickness, and performance characteristics, so it's important to choose the right membrane for the specific application.

12. **Waterproofing is only necessary during construction:** Waterproofing should be an ongoing maintenance consideration for buildings to ensure continued protection against water damage during service life.

13. **Waterproofing ruins the aesthetics of a building:** There are many waterproofing products and systems available that are designed to be discreet and preserve the appearance of a building.

14. **Waterproofing is only for preventing leaks:** While preventing leaks is an important function of waterproofing, it can also protect structures from water damage, mould and corrosion, also longevity of structure.

15. **Waterproofing is only necessary for buildings near water bodies:** While buildings near water bodies are more susceptible to water damage, all buildings can benefit from waterproofing to protect against leaks and moisture intrusion.

16. **Waterproofing makes buildings completely impervious to water:** While waterproofing can significantly reduce the risk of water damage, it may not make a building completely impervious to water under all conditions.

17. **Waterproofing is only for preventing external water ingress:** Waterproofing can also prevent water vapour

from entering a structure, which can cause condensation and mould growth.

18. **Waterproofing is only necessary for buildings with flat roofs:** While flat roofs are more prone to water pooling and leaks, waterproofing is important for all types of roofs to prevent water damage for protecting structure.

19. **Waterproofing is Harmful to the Environment:** All waterproofing materials are not harmful to the environment. Many Eco-friendly waterproofing products are now available in the market.

20. **Waterproofing is a DIY (Do it yourself) job:** While some minor waterproofing tasks can be done by homeowners, complex waterproofing projects often require professional expertise.

21. **Waterproofing is unnecessary if a building has good drainage:** While proper drainage is important, waterproofing provides an additional layer of protection against water damage.

22. **Waterproofing is unnecessary if a building has a sloped roof:** While sloped roofs help to direct water away from the building, waterproofing is still necessary to prevent leaks and water damage.

23. **Waterproofing paint is sufficient for prevent water ingress:** Waterproofing paint cannot prevent water

ingress. Paint can provide a temporary barrier but isn't a long-term solution for structural waterproofing.

24. **Waterproofing is a one-size-fits-all solution**: Different buildings have different waterproofing needs based on factors such as location, climate, and construction materials.

25. **Waterproofing can fix structural issues**: While waterproofing can prevent water damage, it cannot fix underlying structural issues that may be causing leaks or moisture intrusion.

Please scan the QR Code for further information:

"Our greatest weakness lies in giving up. The most certain way to succeed is always to try just one more time."

-Thomas Edison

Chapter

3

Main Problems/ Issues Faces in Waterproofing

Waterproofing is crucial in construction to prevent water ingress, which can lead to structural damage, mould growth, and many other issues. Here are 15 common problems or issues faced in waterproofing work:

1. **Cracking:** Settlement, structural movement or improper installation can cause cracks in the waterproofing membrane, allowing water to seep through.
2. **Joint Failure:** Improperly sealed joints between different building elements can allow water to infiltrate.
3. **Improper Surface Preparation:** Poor surface preparation can compromise the adhesion of waterproofing materials, leading to leaks.
4. **Inadequate Material Selection:** Choosing the wrong waterproofing material for the specific application or environment can result in premature failure.
5. **Inadequate Coverage:** Insufficient application of waterproofing materials can leave vulnerable areas exposed to water penetration.
6. **Poor Installation Practices:** Incorrect installation techniques, such as inadequate curing times or improper

seam sealing, can compromise the waterproofing system.

7. **Inadequate Sloping or Drainage:** Poor design or installation of drainage systems can lead to ponding water, increasing the risk of leaks.

8. **Inadequate Maintenance:** Lack of regular inspection and maintenance can allow small issues to escalate into larger problems over time.

9. **Incompatible Materials:** Compatibility issues between waterproofing materials and substrates can lead to adhesion failure and leaks.

10. **Chemical Damage:** Exposure to chemicals, such as those found in soil or certain construction materials, can degrade waterproofing membranes over time.

11. **UV Damage:** UV radiation can deteriorate certain types of waterproofing materials exposed to sunlight, leading to reduced effectiveness.

12. **Poor Design:** Inadequate consideration of waterproofing requirements during the design phase can result in inadequate protection against water intrusion.

13. **Temperature Variations:** Extreme temperature fluctuations can cause expansion and contraction of building materials, potentially compromising waterproofing integrity.

14. **Water Pressure:** High hydrostatic pressure, such as that found in basements or below-grade structures & also overhead water body or swimming pool, etc. can challenge waterproofing systems.

15. **Biological Growth:** Mold, algae, or other biological growth can compromise the integrity of waterproofing materials and contribute to water penetration.

Addressing these issues requires careful planning, proper material selection, skilled installation, waterproofing consultant supervision and regular maintenance to ensure effective waterproofing performance over time.

Please scan the QR Code for further information:

"Let's not get tired of doing good, because in time we'll have a harvest if we don't give up."

-Bible Verses

Chapter

4

Waterproofing: - Need of Today

4.1 Water Proofing FAQS (Frequently Asked Questions)

Q-1: What is waterproofing?

A-1: Waterproofing is such a Coating System through which water can not enter in the Structure/Mother Surface of the Construction.

Q-2: Why was waterproofing not necessary or little necessary in earlier times?

A-2: Before 15-20 years, waterproofing was not required due to the following reasons:

1. At that time the exterior walls of residential/commercial complexes were 9" in size and most electrical wiring was not even concealed. Due to this, the size of the crack leakage in the outer wall was minute & not much and therefore the rain water was not in a situation to enter the wall.
2. While constructing usually, D.P.C. (Damp proof course) was done on the inner and outer wall of the building. Therefore, rising dampness or capillary action was in a position to stop humidity/dampness.

3. The changes in climate were also not much compared to the current situation.

4. In earlier times, labourers used to provide quality work while doing construction as compared to the present day.

5. The work was completed very peacefully without any haste by the labourers. So, the quality of construction was good.

Due to the above reasons, there was little requirement for waterproofing materials in earlier times, and hence, the importance of waterproofing was not high.

Q-3: Why is waterproofing necessary/ must in today's time?

A-3: Water Proofing is necessary for today's time because:

1. In today's construction, building exterior walls are 4.5" thick instead of 9" thick and electrical wiring is concealed rather than exposed. Additionally, the outlook/elevation/groove of the building improved in the process of making the building attractive, and hence, there are much more chances of water dampness/humidity. Therefore, waterproofing is necessary to maintain the beautiful elevation of building & to protect the thin wall from water seepage.
2. In most of the constructed buildings today, the external and internal walls are of 4.5″ size and in

some places, the walls are made of cement blocks (ACC blocks) and this leads to dampness in the wall and infiltration of rainwater from outer side.

3. In today's practice, there is no RCC sill band & lintel band, So further expansion & contraction is taking place as the time passes on, which result in cracks. So, water can easily penetrate from the wall area causes humidity/dampness/seepage of water.

4. Due to global warming, the heat is increasing in the summer time, causing cracks in the walls and creating a water seepage problem.

5. At present, there is a lack of labours and their efficiency is not much high, it affects the quality of work. Thereafter, the owner has to face the problem of water.

6. At this time, everyone has to speed up to complete the work because the contractor is given a specific time target to complete the work (within a certain time limit). Due to this, the quality of work is affected, resulting in seepage/leakage of water.

7. People who purchase a residence or office spend a lot of money; some of their whole lives are invested in such a purchase, and a lot of money is spent on interior decoration. In these circumstances, if the water problem arises later, the buyer will suffer a huge loss. To prevent loss, it is necessary to get waterproofing

of any residence/office/complex under the main supervision of a waterproofing expert. In addition, the cost of waterproofing is very insignificant compared to the total cost of the house/ office/ complex, etc.

Please scan the QR Code for further information:

"Experience is a hard teacher because she gives the test first, the lesson afterwards."

- Vernon Sanders Law

Q-4: What is the difference between Basic Water Proofing and Quality Water Proofing?

A-4: Basic Water Proofing: -

1. Self-waterproofing with the basic economic waterproofing product available in the market.
2. Using a same waterproofing product for different area

of waterproofing instead of using the suitable waterproofing product for different areas.

3. Waterproofing work will be assigned to a general plumber or local waterproofing agency instead of a waterproofing specialist.

If this type of waterproofing is done, it will give an initial level of satisfaction and it is also cheap. However, for any reason, if water drainage continues, the initial expenses will be wasted and eventually become a costly affair.

Quality Water Proofing: -

1. Selection of the best waterproofing products (for different areas of waterproofing) from the number of products available in the market with the minimum cost.

2. A specialist waterproofing consultant will also explain why to use a certain product and not others and how to use it to get maximum results.

3. Along with his expertise, a waterproofing specialist also hires technical staff to give maximum satisfaction to his clients (because waterproofing is a highly technical job) and therefore, the cost incurred for the waterproofing

work will be the resulting cost and it will increase brand value of house/shop/building/complex/industry.

In short, Quality waterproofing means

1. Saving of interior decoration from water seepage
2. Health-related problems are also solved
3. Saving costs from other expenses
4. Long term relief

Q-5: Waterproofing is to be done in which area?

A-5: In the following areas of the building, waterproofing is necessary.

i. Structure of the building (during construction).

ii. Areas from which rainwater is likely to enter in the building.

iii. Toilets/Bathrooms/Water tank/ Podium - Landscape - Hardscape area of the building from where there is a possibility of water seepage.

For example,

In Residence or Commercial buildings, Waterproofing at the following places is a must.

1. Retaining/ Diaphragm Wall & Raft Foundation
2. Toilets/Bathrooms
3. Kitchen/Wash

4. Balcony/Roofs/Terrace/Chhajja/Planters/Water Tank area
5. Waterbody/Swimming pool/ Garden (Podium) area
6. Parking area
7. External wall area of the building, etc.

Please scan the QR Code for further information:

"The moment we believe that success is determined by an ingrained level of ability as opposed to resilience and hard work, we will be brittle in the face of adversity."

— Joshua Waitzkin

Q-6: Which points should be kept in mind while going for waterproofing?

A-6: The following points to be considered while carrying out waterproofing work

i. In today's market, many Water Proofing products are available. A Waterproofing Consultant needs to be consulted about the application of a particular product for a particular area and only then should the waterproofing work to be started.

ii. Coordination between the interior decorator/architect/ consultant and the waterproofing consultant/Expert is necessary so that they all decide on a joint work plan for best results by finalising the final finished product before starting the waterproofing work.

iii. Waterproofing work to commence only after completion of concealed electrical wiring/plumbing work/A.C. pipefitting, etc.

iv. The area where the waterproofing work will be carried out will be cleaned by the concerned cleaning agency (before starting the waterproofing work).

v. A Water proofing specialist Contractor will now inspect the area and start preparing the entire surface to carry out the work.

vi. A waterproofing specialist Contractor will now ensure micro and deep cleaning of the waterproofing area.

vii. The Loose particles in the waterproofing area will now be cleaned, after opening the cracks (grooves), will be filled with crack filler or Polymer Modified Mortar (PMM) under Water proofing consultant supervision as required.

Please scan the QR Code for further information:

"I never dreamt of success. I worked for it."

-Estée Lauder

viii. To know the strength of the existing structure and type of waterproofing, a water test (Pond Test) needs to be conducted and if leakage is found during the pond test, Cementitious Injection Grouting (Positive or Negative side) and if needed. In addition, if the need arises, higher-grade material should also be used to get satisfactory results for a longer period.

ix. Now the boreholes (drainage/down take pipes holes/ Water line pipe holes) in the area of waterproofing should also be taken care of, normal shuttering/ non-shrink free flow cementitious grout should be applied, and if possible, Using Epoxy Material or as a Combination Separately further P.U. sealant and fibre net/tape joint system. Further, any borehole should also be pond tested and if any leakage is found, it will be repaired and waterproofing coating will be applied.

x. While carrying out water proofing, there is a need to fill Polymer Modified Mortar (PMM) in the ratio of (SBR: Cement: Water) :(1:1:3) in the floor and wall area or Do in Ratio as per TDS (Technical Data Sheet)/PDS (Product Data Sheet) with consultation of waterproofing consultant's specification or guideline and thereafter fibre net Sandwich Vata of 2" to 4" or with As per site situation to be done and on 2nd day Cement Mortar Vata to be done after consulting tiles agency and site in-charge, so that in future there is no need of its re-opening.

xi. At this stage, waterproofing in the decided area will be started with the decided specific product. First of all, the area of waterproofing to be made SSD (Surface Saturated Dry) as per product TDS/PDS. Now first coat of water proofing to be applied horizontally. After that as per the advice of water proofing consultant & As per Technical Data Sheet (TDS), vertical coat of water proofing to be done. Now as per the necessity need to use fibre net or non-woven geo-textile and increase the additional coating as per requirement.

xii. Generally, the thickness of the waterproofing coating is upto 1 to 2 mm, and to prevent damage to the waterproofing membrane, So, if there is no stagnant water in specified area during monsoon or while day to day use. So no leakage occurs during service life. there is a need of speedy plaster/ putty with adding integral waterproofing compound as per site condition for wall area & for floor area Brick Bats Cube Concrete/India Waterproofing (IWP) or IPS or Screeding with adding integral Waterproofing Compound is to be done as per site condition by using proper slope if possible 1" for 10'-0" distance. So that water does not accumulate and waterproofing work gives proper results during the service cycle.

xiii. After completion of above process, the waterproofing area needs to be watered manually for at least three (3) days to maximum seven (7) days and after verification, further floors or finishing work is to be undertaken.

If above mentioned points are kept in mind while doing water proofing, we will surely that in a position to make our goal of Leak Free and Healthy India.

Please scan the QR Code for further information:

"Goal setting is the secret to a compelling future."

- Tony Robbins

4.2 Waterproofing at Various Areas

Generally, waterproofing is required in all areas where there is a chance of rainwater or Tank water from the Bathroom or leakage/moisture from outside, entering the structure/ building in the area.

Waterproofing should be done at retaining wall, building external wall area, bathroom, kitchen wash, balcony, porch, terrace, water tank, water body, swimming pool/lily pond, garden, planters, fountain, podium area, parking area.

1. **Retaining wall/Diaphragm wall**: Waterproofing of retaining wall/diaphragm wall is essential to avoid serious damage to the building structure and to increase the service life of building.

2. **Raft foundation**: Raft foundation is the foundation of the structure, to support the structure for its utility purpose. It's Waterproofing is very important and needs to be done with accuracy.

3. **Basement**: Basement is a part of the foundation structure of your entire building. This simple fact is enough to justify waterproofing of basement area.

4. **Lift-Pits**: As an integral part of the structure, lift pits plays an important role in carrying and bearing significant loads. Hence, the lift pit must be waterproofed in today's senerio.

5. **Expansion Joints:** Expansion joints are installed to manage the co-efficient of thermal expansion over large spans of concrete. For that reason, the waterproofing material must not only withstand the movement of the building but must also be designed to resist ageing, which leads to the weakening of the bond on the concrete and causes the failure of the system.

6. **Underground Water Tank (UGWT):** Underground water tanks are insecure from external sources such as surrounding soil and soluble salts and, if not properly waterproofed, can cause attacks on the reinforcement after it enters the tank.

7. **Landscape/Garden:** Landscape/Garden waterproofing is one of the basic construction requirements. The Waterproofing coating should be anti-root, and this area should be in contact with water for the maximum amount of time. So, the waterproofing coating should be such that it can transmit the water bar with proper drainage system.

8. **Hardscape/Driveway:** Hardscape/Driveway covers a large area so its waterproofing is very important. Thus, it should be waterproofed with proper slope and screed/IPS.

9. **Podium/Parking Area:** Waterproofing coating of

Podium/Parking area should have adequate flexibility and flexural strength. And Minor/Major cracks should be treated properly while waterproofing work.

10. **Toilet/Bathroom**: Toilet/Bathroom are exposed to water for a long period of time. The waterproofing coating should have good flexibility and covered with proper slope of screed/Brick Bats Cuba Concrete/ IWP/IPS.

11. **Chhajjas/Planter:** The exterior sides of the building such as roof and planter area should be properly waterproofed. You can waterproof the eaves with a suitable waterproofing coating and complete it with a slope. The waterproofing coating should be anti-root for planters and since the area will be exposed to water for maximum period of time, the waterproofing coating should be able to withstand the flow of water.

12. **Terrace Waterproofing**: While waterproofing of terrace, you should apply a thermal insulating or heat reflective coating with the waterproofing

13. **Overhead Water Tank (OHWT)**: overhead water tank waterproofing is very very important to prevent water leakage, to maintain structural longevity, for CFTRI approved food grade purpose & also protect Algae & mold growth for keeping its durability for it's service life.

14. **Water Repellent Coating**: An acrylic-based coating can be applied to the outer/exterior wall followed by a silicone-based water-repellent coating, for making entire building as a waterproofing membrane shield.

Please scan the QR Code for further information:

"Either you run the day or the day runs you."

- Jim Rohn

4.3 Admixture at the Time of Casting

Admixtures are additives to the concrete mixing of it can help to control the setting time and other aspects of fresh concrete.

Admixtures are used for water reduction, slump reduction, strength gain, high workability, stickiness of concrete, wear resistance and high finish for modern structures.

Admixture constituents by volume as below,

Coarse Aggregate 43 %, Sand 33%, Cement 12%, 5% SCM, 5-7 % Water, 1-2 % Air, < 0.2 % Admixture

In General, admixtures portion is minimal part of the concrete but it is very effectively as compared to the other concrete parts. Therefore, Admixtures are essential components of the concrete.

As per present scenario, concrete is used with high workability, strength, sustainability, smaller or slender elements best suitable for compatible structures.

In Today's modern scenario, builders will prepare a high-rise building. So, Admixtures will be used as eco-friendly, extremely strong & durable, smart unmatched workability & aesthetics, tailor made as well as insulation & light weight conductivity, where we will used M170 grade concrete. Thus, India is also improving admixtures quality day by day with latest technology.

Types of admixtures:

- Accelerating Admixtures
- Retarding Admixtures
- Air-Entraining Admixtures
- Water Reducing Admixtures
- Superplasticisers
- Waterproofing admixture
- Pozzolanic or Mineral Admixtures

Material used as Integral Waterproofing Compound:

In conventional materials, admixtures are used as water reducers, air-entraining admixtures, and polymer-based admixtures. Water reducers are mainly used Lignosulphates , SNF, SMF, and PCE bases. Air entraining is mainly used on soap, Rosinates , AOS etc. bases. Polymer-based admixture is mainly an Acrylic/SBR (Styrene Butadiene Rubber) emulsion.

Generally, Fatty acid materials have been used in older times. But as per the current scenario, there are different types of material used i.e. Lignosulphates , SMF, SNF, PCE, etc.

Lignosulphates based materials have been used for M10 to M30 grade concrete, where its dosage = 0.2 to 0.6%.

SMF-based materials have been used for M20 to M50 grade

concrete, where its dosage = 1.0 to 3.0%.

SNF-based materials have been used for M20 to M50 grade concrete, where its dosage = 0.8 to 0.6%.

PCE-based materials have been used for M30 to M80 grade concrete, where its dosage = 0.3 to 3.0%.

Hence, Integral waterproofing additives have been used for plasters, concrete, RMS, precast concrete as well as facade walls, etc.

Please scan the QR Code for further information:

"Success is getting what you want;
happiness is wanting what you get."

-W. P. Kinsella

4.4 Health & Safety aspects in waterproofing works

The Significance of Health and Safety in Construction

Health & Safety in construction cannot be avoided under any circumstances, as there are many hazards and chances of accidents.

In the construction industry, Waterproofing activities are full of risk and, hence, require utmost attention.

Certain confined spaces include Manholes, Septic tank (STP), Storage Tanks, Pits, Retaining Wall, and Swimming Pool etc. So, proper precautions have to be taken before anyone can take entry in it. After making arrangement of Proper ventilation, the workers should take entry in this area.

Actually, Personal Protective Equipment (PPE) are to be ensured when carrying out waterproofing work in the confined spaces mentioned above.

The following PPEs are required for the locations mentioned here in below.

SR NO	LOCATION	REQUIRED PPEs
1	Below Ground Level Area (Such as Underground, Water tank, Septic tank, Basement, Retaining Wall, etc.)	Safety Helmet, Safety Shoes/Gumboot, Rubber Hand Gloves, Dust mask /Face shield, Safety Glasses, Airline mask (If necessary), Safety belt (If necessary), Reflective Jacket, First Aid Kits, Fire Extinguisher
2	Inside Building Waterproofing Area (Such as Bathroom, Kitchen, floors, etc.)	Safety Shoes, Rubber Hand gloves, Safety Glasses/ Face Shield, First Aid Kits & Fire Extinguisher
3	Top Terrace, Over Head Water tank& ESR.	Safety shoes, Safety Helmet, Safety Belt, Safety Rubber Hand gloves, Face Shield, Safety Glasses, First Aid Kits & Fire Extinguisher, Airline mask (As per requirement), Reflective Jacket

For health & safety well define exit plan i.e. Emergency Exit Plans is to be prepared well in advanced & also do regular mock up sessions to ensure readiness.

Thus health & safety is required in waterproofing industry for safety of workers, Protection of Environment, using different types of chemicals for different waterproofing areas, avoiding legal compliance, creating Reputation & Trust, also for quality Assurance, increasing efficiency & productivity also for awareness in using modern chemical materials & tools in our waterproofing industry for creating positive business reputation.

Please scan the QR Code for further information:

"One secret of success in life is for a man to be ready for his opportunity when it comes."

– Benjamin Disraeli

Chapter

5

Sub Structure Waterproofing

5.1 Retaining Wall / Diaphragm Wall Waterproofing

The Waterproofing of Retaining Walls/Diaphragm Walls are very essential & unavoidable for every building who has basement in it. Retaining Walls/Diaphragm Walls are constantly exposed to humidity. So, to avoid serious damage to the structure of the building and major decorative damage to rooms inside if proposed in some structures the waterproofing of Retaining/ Diaphragm walls water proofing is required.

There are many types of Waterproofing Available in Indian Market for Retaining Retaining wall / Diaphragm walls like

- Admixture using as a integral waterproofing part while/ during construction
- Membrane Sheets - like bitumen, PVC or HDPE
- (LAM) Liquid Applied Membranes bituminous Acrylic based, cementitious based, polyurethane based, polyurea based
- Crystalline waterproofing
- SBS (Styrene - Butadiens - Styrene) Membrane
- Bentonite Cement Mixes & Sodium Bentonite sheets etc.

Specification & System of Waterproofing

Step 1 (A) - Cold joint/Junction Area Treatment at Retaining Wall, Epoxy/Cementitious Injection grouting for strengthening at Retaining Wall/ Diaphragm Wall (If required)

Demolition or Removal of existing damaged/broken/loose R.C.C. area. [If required]

- Drill holes with electric drilling machines along the honeycombed area and, if required, at the floor wall joint slab and fix PVC nozzles in place using a quick-setting compound.
- Filling Polymer Modified Mortar in Honey Combed Area after finalising & fixing the nozzles. (if required)
- Applying Epoxy/Cementitious Injection Grouting Work with the grouting machine through the nozzles, starting from one nozzle and stopping the operations when the grout starts coming out from the adjacent nozzle; this is done similarly in all the nozzles. [As per client's suggestion or actual site condition requirement]
- After Completion of Grouting work, Remove or Cut the PVC nozzles & plug the holes filled by Non-Shrink Free Flow Cementitious Grout/ Repair Mortar / Polymer Modified Mortar (PMM).

Step 1 (B) - Cold joint/Junction Area Treatment at Retaining wall/ Diaphragm wall

- Cleaning Work as per requirement.
- Applying Crack filling/PMM work as per Site Conditions.
- Applying a fibre net / non-woven geotextile of a 2 to 4-inch-wide sandwich coat or an imported tape joint system with adhesive at all junction areas.
- Applying Cement mortar vata / Angle fillet vata & PMM as per site conditions

Step 2 - <u>Tie Rod Hole Pack at the Retaining Wall</u>

- Tie Rod Hole Pack with Polyester Resin Grout / Non-Shrink Free Flow Cementations Grout / Repair Mortar / Polymer Modified Mortar (PMM) as per site Condition.

Step 3 – Waterproofing Coating Work

[Cleaning + Primer Coat + Waterproofing Coating - 3 Coats or as per Technical Data Sheet (TDS)]

- Surfaces to be waterproofed shall be sound, clean, free of dust, paint, Curing agents, loose plaster, debris etc.
- All final cleaning will be carried out with an electric blower & Water jet (if required).
- Finally, all the surface areas shall be dampened

thoroughly and cleaned with water, as this is necessary prior to the application of any type of coating.

- Applying Crack filling Work with Crack Filler & Polymer Modified Mortar (PMM) at Junction & Honeycombed Area with SBR: Water: Cement.
- Applying Primer Coat as a 1st layer.
- Applying Waterproofing Coating – 3 coats (as per TDS), coat should be applying perpendicular to the previous one.

Waterproofing Coating Work with 10 Options

Applying Polyurethane Coating with Spray Application or by manual way

OR

Applying PVC membrane

OR

Applying HDPE membrane

OR

Applying Bentonite Cement mixer or Sodium Bentonite sheets

OR

Applying Bituminous Coating – 2 or 3 Coats

OR

Applying Crystalline Coating – 2 or 3 Coats

OR

Applying Two Component Cementitious Coating – 3 Coats

OR

Applying APP Membrane

OR

Applying SBS Membrane with Geotextile

Step-4: Steel Bars Surrounding Hole Pack with Polyester Resin Grout

- Applying Steel Bars Surrounding the Area Hole Pack with Polyester Resin Grout.
- Also providing Additional 4" × 4" - 40 GSM non woven geotextile or Imported Tape Joint with Sandwich Coat as per site conditions.

Step – 5 - Applying Protection Layer

- Applying plaster/PMM to the wall area with an integral waterproofing compound.

OR

Applying Dimple Sheet for Protection layer.

OR

Applying any kind of Protection Layer Sheet,
as per site conditions.

Sr. No.	Description
1	Tape Joint/Fiber Mesh @ Cold/Construction
2	Angle Fillet/ Vata
3	Tie Rod
4	Primer Coat
5	Polyurethane Coating/ PVC or HDPE Membrane/ Bentonite Cement Mixer or Sodium Bentonite Sheets/ Bituminous Coating/ Crystalline Coating/ Two Component Cementitious Coating/ App Membrane / SBS Membrane
6	Protection Layer

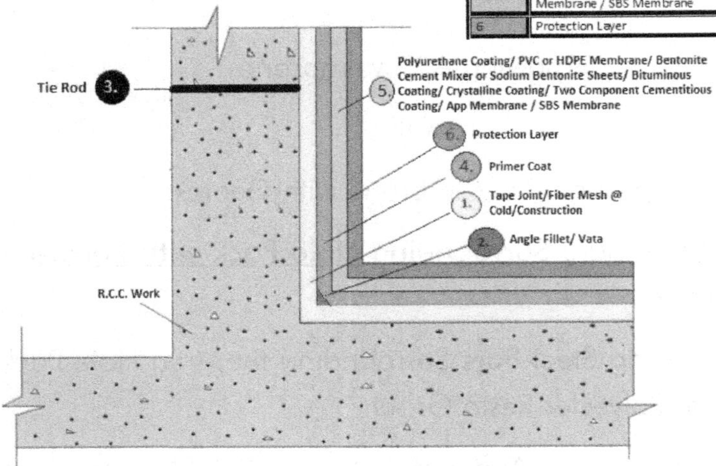

Fig. 5.1 Retaining Wall Waterproofing

Please scan the QR Code for further information:

"Education is the most powerful weapon which you can use to change the world."

- Nelson Mandela

5.2 Raft Foundation Waterproofing

Raft foundation is a thick concrete slab resting on a large area of soil reinforced with steel, supporting columns or walls and transfer loads downwards from the structure to the soil.

Raft foundation is generally used to support structures like residential or commercial buildings where soil condition is poor, storage tanks, silos, foundations for heavy industrial equipment etc.

SPECIFICATION & SYSTEM OF WATERPROOFING

Step-1: Cleaning Work & Crack filling

Surfaces to be waterproofed shall be sound, clean, free of dust, paint, Curing agents, loose plaster debris etc.

All final cleaning will be carried out with an electric blower & Water jet (if required).

Finally, all the surface areas shall be dampened thoroughly with clean water, as this is necessary prior to the application of any type of coating.

Applying Crack filling Work with Crack Filler & Polymer Modified Mortar (PMM) at Junction & Honeycombed Area with SBR: Water: Cement.

Step 2: Construction Joint Area Treatment

- Cleaning Work as per requirement.

- Applying Crack filling/PMM work as per Site Conditions.
- Applying an Imported Tape Joint System with Adhesive at All Junction Areas.

<u>Step-3: Active Leakage Area – Injection Grouting Work (If required) – PU / Cementitious</u>

- Demolition or Removal of existing damaged/broken/ loose R.C.C. area.
- Drilling 14mm diameter holes with electric drilling machines along the honey combed area if required at floor wall joints slab and fixing Aluminium or G.I NRV heavy nozzles of 10 mm diameter in place using quick setting compound.
- Filling Polymer Modified Mortar in Honey Combed R.C.C/Brick Masonry Area after finalising & fixing the Nozzles.
- Pressure grouting should be carried out by using [PU FOAM] Hand operated Centrifugal PU grouting machine through the nozzles, starting from one nozzle and stopping the operations when the grout starts coming out from the adjacent nozzle; this is done similarly in all the nozzles. [As per client's suggestion or actual site condition requirement]
- Pressure grouting should be carried out by using [PU

RESIN], An operated Centrifugal PU grouting machine through the nozzles, starting from one nozzle and stopping the operations when the grout starts coming out from the adjacent nozzle; this is done similarly in all the nozzles. [As per client's suggestion or actual site condition requirement]

- After Completion of PU Grouting work, Remove or Cut the Aluminium or G.I NRV heavy nozzles & plug the holes filled by quick setting compound or Non-Shrink Free Flow Cementitious Grout/ Repair Mortar /Polymer Modified Mortar (PMM).
- Providing & Applying Crystalline Waterproofing -2 Coats on Only the Nozzles surrounded area.
- Make curing or apply curing compound to that area. [If required]

OR

- Applying Cementitious Injection Grouting Work as per Site requirement [If required].

Step-4: Waterproofing Coating Work

- Laying PVC (Poly Vinyl Chloride)/ HDPE (High Density Poly Ethylene)/ EPDM (Ethylene Propylene Diene Monomer)/ EPMS (Ethylene Propylene Monomer System)/ Polyurea Waterproofing/ Polyurethane Waterproofing/ Cementitious Waterproofing/

Crystalline Waterproofing/ Bentonite Waterproofing Combination/ Bituminous Waterproofing

Step-5: Applying Protection Layer

- Applying IPS/Screed on floor area as per Site Conditions.
- Applying plaster/PMM to the wall area with an integral waterproofing compound.

OR

- Applying Any kind of protection layer sheet.

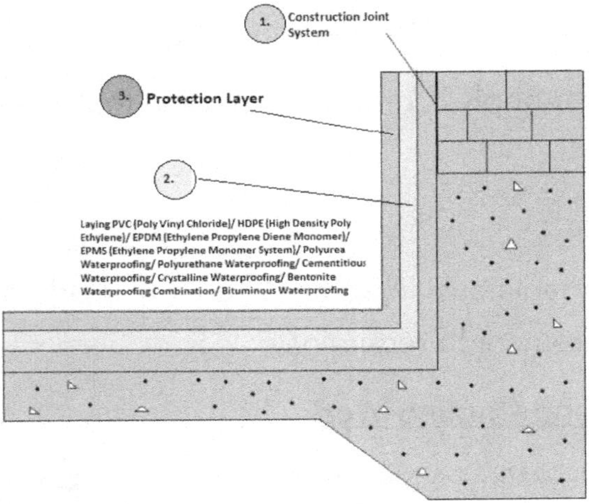

Sr. No.	Description
1	Construction Joint System
2	Laying PVC (Poly Vinyl Chloride)/ HDPE (High Density Poly Ethylene)/ EPDM (Ethylene Propylene Diene Monomer)/ EPMS (Ethylene Propylene Monomer System)/ Polyurea Waterproofing/ Polyurethane Waterproofing/ Cementitious Waterproofing/ Crystalline Waterproofing/ Bentonite Waterproofing Combination/ Bituminous Waterproofing
3	Protection Layer

Fig. 5.2 Raft Foundation Waterproofing

Please scan the QR Code for further information:

"Success is stumbling from failure to failure with no loss of enthusiasm."

-Winston Churchill

5.3 Basement Waterproofing

People use basement area as a place to store wastes & non useful goods. It's the foundational structure of the building.

Basement waterproofing is required for service life of structure, healthy & safely living of occupant , protecting water leakage & air quality.

It is very essential to make sure that water does not accumulate around it.

SPECIFICATION & SYSTEM OF WATERPROOFING

Step-1: Cleaning Work & Crack filling

Surfaces to be waterproofed shall be sound, clean, free of dust, paint, Curing agents, loose plaster debris etc.

All final cleaning will be carried out with an electric blower & Water jet (if required).

Finally, all the surface areas shall be dampened thoroughly with clean water, as this is necessary prior to the application of any type of coating.

Applying Crack filling Work with Crack Filler & Polymer Modified Mortar (PMM) at Junction & Honeycombed Area with SBR: Water: Cement.

Step 2: Construction/Expansion Joint Area Treatment

- Cleaning Work as per requirement.
- Applying Crack filling/PMM work as per Site

Conditions. (Refer Chapter - 7.2)

- Applying an Imported Tape Joint System/ Fiber Net/ Non-woven Geotextile with Adhesive at All Junction Areas.

Step-3: Active Leakage Area – Injection Grouting Work (If required) – PU / Cementitious

- Do naked pond test on basement area stage by stage as per site conditions, then check it out from negative side that is bottom side or basement area ceiling side of leakage/seepage or dampness appear, Do start PU Injection grouting from negative side.
- Demolition or Removal of existing damaged/broken/ loose R.C.C. area.
- Drilling 14mm diameter holes with electric drilling machines along the honey combed area and if required at floor wall joints slab and fixing Aluminium or G.I NRV heavy nozzles of 10 mm diameter in place using quick setting compound.
- Filling Polymer Modified Mortar in Honey Combed R.C.C/Brick Masonry Area after finalising & fixing the Nozzles.
- Pressure grouting should be carried out by using [PU FOAM] Hand operated Centrifugal PU grouting machine through the nozzles, starting from one nozzle and stopping the operations when the grout starts

coming out from the adjacent nozzle; this is done similarly in all the nozzles. [As per client's suggestion or actual site condition requirement]

- Pressure grouting should be carried out by using [PU RESIN], An operated Centrifugal PU grouting machine through the nozzles, starting from one nozzle and stopping the operations when the grout starts coming out from the adjacent nozzle; this is done similarly in all the nozzles. [As per client's suggestion or actual site condition requirement]
- After Completion of PU Grouting work, Remove or Cut the Aluminium or G.I NRV heavy nozzles & plug the holes filled by quick setting compound or Non-Shrink Free Flow Cementitious Grout/ Repair Mortar /Polymer Modified Mortar (PMM).
- Providing & Applying Crystalline Waterproofing -2 Coats on Only nearby the Nozzles surrounded area.
- Make curing or apply curing compound to that area. [If required]

OR

- Applying Cementitious Injection Grouting Work as per Site. [If required]

Step-4: Waterproofing Coating work with 10 Options

Laying APP (Atactic Poly Propylene) or SBS (Styrene

Butadiene Styrene) membrane

OR

Applying Bituminous Coating – 2 or 3 Coats

OR

Applying Crystalline Coating – 2 or 3 Coats

OR

Applying Two Component Cementitious – 3 Coats

OR

Applying HDPE/PVC/EPDM Membrane with Geotextile

OR

Applying Polyurethane Coating

OR

Applying Polyurea Waterproofing Coating

Step-5: Applying India Waterproofing (IWP)/Brick Bats Cuba Concrete (B.B.C.C.)/ IPS/Screeding on Floor Area & Plaster/PMM on Wall as per site requirement

- Apply Geo-textile/Fiber Net on the Floor + Apply IWP/IPS/Screeding on the Floor Area with Integral Waterproofing Compound.
- Apply plaster/PMM on the wall area with an integral waterproofing compound.

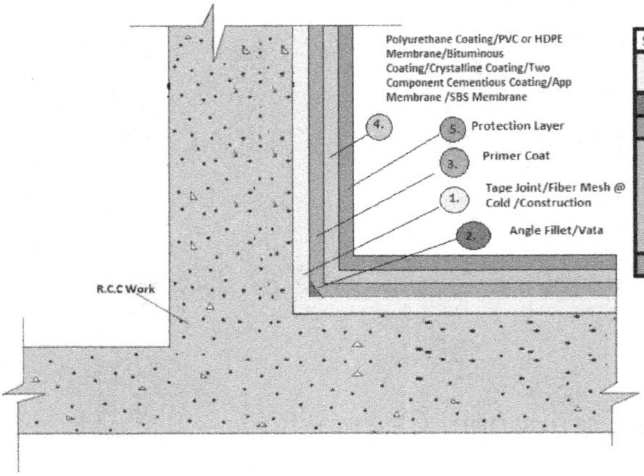

Sr. No.	Description
1	Tape Joint/Fiber Mesh @ Cold/Construction
2	Angle Fillet/ Vata
3	Primer Coat
4	Polyurethane Coating/PVC or HDPE Membrane/Bituminous Coating/Crystalline Coating/Two Component Cementious Coating/App Membrane /SBS Membrane
5	Protection Layer

Fig. 5.3 Basement Waterproofing

Please scan the QR Code for further information:

"Optimism is the faith that leads to achievement. Nothing can be done without hope and confidence."

- Helen Keller

5.4 Lift Pit Waterproofing

Lift Pits as an integral part of the structure have an important role to play and a considerable load to bear.

When there is a buildup of contained water in a pit, it mixes with oil and trash, creating odour and putting the shaft integrity at major risk. The disaster could be immediate or gradual– every case is different, but the outcome is usually the same: the reinforcing bar that sustains the wall structure starts to show rusting due to water deteriorating the wall, damage to pit lights and electrical wires, rusted parts, foul odour, mosquitoes and failed inspections.

Thus, in the modern Senerio everyone and particularly Senior Citizens are using lift regularly for high rise buildings/ projects/structure. Lift in unoperational condition is not good in any case. Proper waterproofing in lift-pit area is most essential. So taking of proper care, perfect and highly systematic waterproofing is today's requirement.

5.4.1. Lift Pit : Inside Area

SPECIFICATION & SYSTEM OF WATERPROOFINGG

Step-1: Applying PU Injection Grouting Work in lift pit entire area as per site conditions.

- Demolition or Removal of existing damaged/ broken/ loose R.C.C. area. (if required)
- Drilling 14mm diameter holes with electric drilling machines along the honeycombed area and, if required, at floor wall joints slab and fixing Aluminium or G.I. heavy nozzles of 10 mm diameter in place using Quick setting, cement-based, water leakage plugging mortar.
- Filling Polymer Modified Mortar in Honey Combed R.C.C/Brick Masonry Area after finalising & fixing the nozzles. (if required)
- Pressure grouting should be carried out by using [PU FORM] An operated PU grouting machine through the nozzles, starting from one nozzle and stopping the operations when the grout starts coming out from the adjacent nozzle; this is done similarly in all the nozzles. [As per actual site condition requirement]
- Pressure grouting should be carried out by using [PU Resins], an operated PU grouting machine through the nozzles, starting from one nozzle and stopping

the operations when the grout starts coming out from the adjacent nozzle; this is done similarly in all the nozzles. [As per actual site condition requirement]

- After Completion of PU Grouting work, Remove or Cut the Aluminium or G.I. heavy nozzles & plug the holes filled by Quick setting, cement based, water leakage plugging mortar or Non-Shrink Free Flow Cementitious Grout/ Repair Mortar /Polymer Modified Mortar (PMM).
- Applying Crystalline Waterproof coating -2 Coats on surrounding of only the Nozzles area.
- Make curing or apply curing compound to that area. [If required]

<u>Step 2: - Lift Pit Area Waterproofing System</u>

(Cleaning + Imported Tape Vata + Mortar Vata + Primer + Two Component Flexible Crack Bridging Waterproof Coating + Fiber Mesh + Two Component Flexible Crack Bridging Waterproof Coating - 2 Coats)

Methodology:

The lift Pit Area to be waterproofed must be cleaned by the client as RCC & another area in the Wall & Junction area need to be Waterproofed.

[Basic cleaning is done by the Client as per the satisfaction & requirement of the contractor/waterproofing Agency]

Surfaces to be waterproofed shall be sound, clean, free of dust, paint, Curing agents, loose plaster debris etc.

All final cleaning will be carried out with an electric blower & Water jet (if required).

Finally, all the surface areas shall be dampened thoroughly with clean water, as this is necessary prior to the application of any type of coating.

- Applying Polymer Modified Mortar (PMM) at Junction & Honeycombed Area With SBR: Water: Cement.
- Applying Imported Tape Vata on Junctions with sandwich coat of two component flexible coating.
- Applying Mortar Vata.
- Applying 1st Coat of Acrylic emulsion Concrete Bonding Agent/ SBR/ Latex or Equivalent as a Primer Coat as a 1st layer.
- Applying 1st Coat of Two Component Flexible Crack Bridging Waterproof Coating or Equivalent as a 2nd layer.
- Applying Fibre Net/Mesh On the previous Coat as a 3rd layer.
- Applying 2nd Coat of Two Component Flexible Crack Bridging Waterproof Coating or Equivalent as a 4th layer.

- Applying 3rd Coat of Two Component Flexible Crack Bridging Waterproof Coating or Equivalent as a 5th layer after 6 to 8 hours of drying of 2nd Coat, as per weather & site conditions.

Step-3: Screeding/Covering Work (Brick Bats Coba Concrete)/India Waterproofing (IWP)/IPS/Plaster /PMM (Polymer Modified Mortar):

- Applying Geotextile after Coating Work & Before Brick Bats Coba Concrete/India Waterproofing (IWP)/ IPS/ Plaster/ PMM (Polymer Modified Mortar) as per site condition.
- Applying India Water Proofing (IWP)/I. P. S. / Screeding on floor area, Applying PMM (Polymer Modified Mortar) Or Plaster on wall area with Integral Water Proofing Compound.
- After Protection Layer any kind of hammering/drilling for installation from lift agency is done then it must be waterproofing again as per site situations, before operation of lift.

Sr No:	Description
1a.	Tape Joint Work
1b.	Cement mortar Vata
2	Primer
3	Two Component 1st Coat
4	Fiber Mesh
5	Two Component 2 Coats
6	Protection Layer

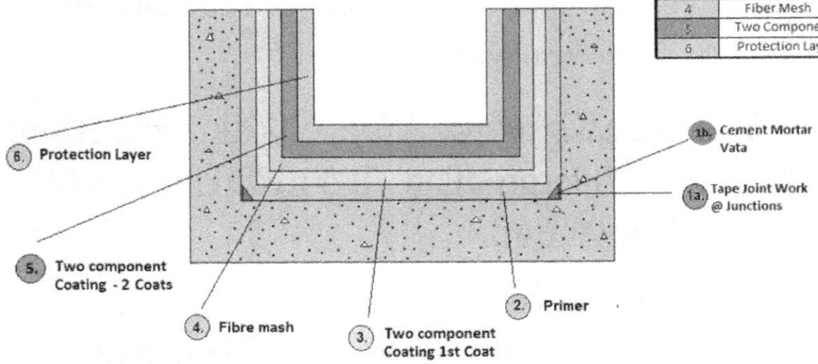

Fig. 5.4.1 Lift Pit Waterproofing: inside Area

5.4.2 Lift Pit :Outside area

Step-1: Outer Side Vata by UV Stable Acrylic Heavy Duty Waterproof Coating System

- Applying Cleaning Work.
- Applying Crack Filler in Cracks.
- Applying Net Vata.
- Applying Mortar Vata.
- Applying Self Penetrating Primer Coat as 1st layer.
- Applying UV Stable Acrylic Heavy Duty Waterproof Coating as a 1st coat as a second layer.
- Applying Fiber Net as a 3rd layer.
- Applying UV Stable Acrylic Heavy Duty Waterproof Coating as a 2nd coat as a 4th layer.
- Applying UV Stable Acrylic Heavy Duty Waterproof Coating as a 3rd coat as a 5th & final layer.

Step – 2- Applying India Waterproofing (IWP)/IPS/ Screeding area as per site requirement

Apply Geotextile/Fibre Net on the Floor + Apply IWP/IPS/ Screeding on the Floor Area with Integral Waterproofing Compound.

Applying PMM(Polymer Modified Mortar) or Plaster on wall area with integral Waterproofing Compound

Sr. No.	Description
1	Net Vata
2	Mortar Vata
3	Primer Coat
4	UV Stable Acrylic Coating 1st Coat
5	Fiber Net
6	UV Stable Acrylic Coating - 2 Coats
7	Protection Layer

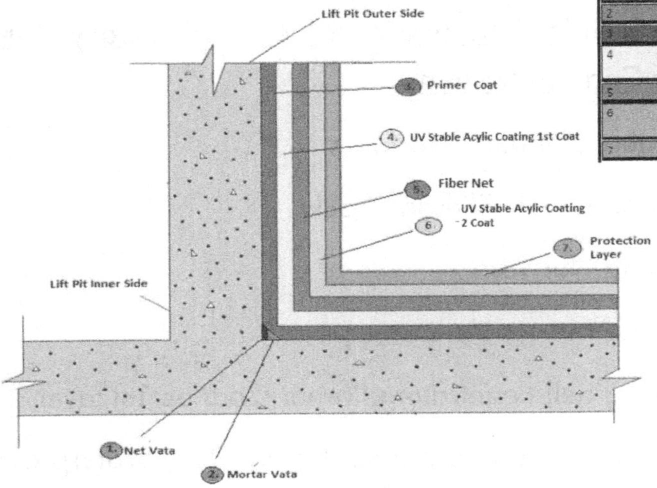

Fig. 5.4.2 Lift Pit Waterproofing: Outside Area

Please scan the QR Code for further information:

"Perfection is not attainable. But if we chase perfection, we can catch excellence."

- Vince Lombardi.

5.5 Construction Joint / Cold Joint / Expansion Joint Waterproofing

There are basically four types of joints based on their Function and construction features. They are listed as,

1. Construction joint
2. Expansion joint
3. Contraction joint/ Control Joint
4. Cold Joint

1. Construction joint

In the concrete structure, these joints are designed to split the large quantum of concrete mass work into a divisional pattern.

2. Expansion joint

Expansion joints are installed to manage the coefficient of thermal expansion over a large span of concrete. The waterproofing material must not only cope with the movement of the building but also be designed to resist ageing, which leads to the weakening of the bond on the concrete and causing failure of the system.

3. Contraction joint / Control Joint

Contraction joints are the joints provided to prevent excessive cracks due to shrinkage and contraction of concrete material. These joints are also called control joints as they help to control the excessive, unnecessary cracks in concrete.

4. Cold Joint

A cold joint is a plane of weakness in concrete caused by an interruption or delay in the concreting operations. It occurs when the first batch of concrete has begun to set before the next batch is added so that the two batches do not intermix. Sometimes, cold joints occur because of emergency interruptions and delays and sometimes because of the work stoppage at the end of the day, but they can also occur from poor consolidation.

Specification: Expansion Joint System of Waterproofing

Step-1: Applying PU sealant With HD 100 Board & Levelling if required (As per Site Condition) (One Part Low Modulus Polyurethane Sealant) or equivalent (PU Sealant about Mother Surface Level)

- Removal / Dismantling of old existing Kotastone/ Granite up to Mother surface, if required.
- Applying Basic Cleaning of Expansion joint & surrounding area of expansion joint.
- Apply grinding of expansion joint surrounding area surface if required.
- Applying edge Repair of expansion joint by cementitious high-grade mortar or equivalent.(If required)

- Applying the Filler board of - Dura Board HD 100 in the expansion joint area, etc., should be complete as per site conditions.(if required)
- Applying PU sealant as per site conditions.

Step 2: Applying Imported Sealing tape with Cementitious Adhesive – Highly elastomeric crack-bridging waterproofing coating

OR

High-Performance Expansion Joint Sealing Tape with Epoxy Adhesive Two Components Resin-Based, Thixotropic, Multipurpose Adhesive

(Tape Joint System about Mother Surface Level)

- Applying Basic Cleaning of Expansion joint & surrounding area of expansion joint.
- Grinding of expansion joint surrounding area surface is applied if required.
- Applying edge Repair of expansion joint by cementitious high-grade mortar or equivalent.(if required)
- Applying Filler board of - Dura Board HD 100 - Make by Supreme in expansion joint area, etc, complete as per site conditions.(if required)
- Applying Highly elastomeric crack bridging

waterproofing coating/Two Component Resin-Based, Thixotropic, Multipurpose Adhesive First Coat as a Primer Coat for Applying tape about the expansion joint and both sides surrounding area.

- Applying Imported Sealing tape / High Performance Expansion Joint Sealing Tape or Equivalent on primer coat about expansion joint and both side surround area.
- Applying highly elastomeric crack bridging water proofing coating/Two Component Resin-Based, Thixotropic, Multipurpose Adhesive First Coat, Second Coat on tape about only both side surround area, keeping expansion joint area without coating to make it expand.
- Applying highly elastomeric crack bridging waterproofing coating on Imported Sealing tape Third & Final Coat on tape about only both sides of the surrounding area, keeping the expansion joint area without coating to make it expand.

Note: About Finish Level Keep Expansion Joint open & apply as per requirement (IPS/Grout/IWP/Tiling/China).

Expansion joint Treatment

(Tape Joint System about Mother Surface Level)

Sr. No.	Description
1	Filler Board
2	PU Sealant
3	Two Component Cementitious or Epoxy Adhesive – 1st Coat
4	Expansion Joint Sealing Tape
5	Two Component Cementitious or Epoxy Adhesive – 2 Coats
6	Protection Cover (GI Sheet/ Aluminium Sheet/ACP Sheet)

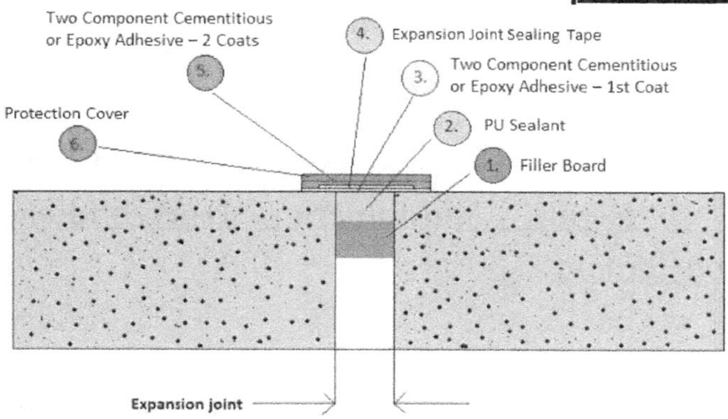

Fig. 5.5.1 Expansion Joint Treatment Tape Joint System at Mother Surface Level

Step-3: Applying PU sealant with HD 100 Board & Levelling if required (As per Site Condition) (One Part Low Modulus Polyurethane Sealant) or equivalent (About Final Finish Level)

- Applying Basic Cleaning of Expansion joint & surrounding area of expansion joint.
- Grinding of expansion joint surrounding area surface is applied if required.

- Applying edge Repair of expansion joint by cementitious high-grade mortar or equivalent.(If required)
- Applying the filler board of Dura Board HD 100 in the expansion joint area, etc. should be completed as per site conditions.(If required)
- Applying PU sealant as per site conditions.

<u>Step-4: Applying Imported Sealing tape with Cementitious Adhesive - Highly elastomeric crack-bridging waterproofing coating</u>

<u>OR</u>

<u>High-Performance Expansion Joint Sealing Tape with Epoxy Adhesive Two Components Resin-Based, Thixotropic, Multipurpose Adhesive</u>

<u>(Tape Joint System about Final Finish Level)</u>

- Applying Basic Cleaning of Expansion joint & surrounding area of expansion joint.
- Grinding of expansion joint surrounding area surface is applied if required.
- Applying edge Repair of expansion joint by cementitious high-grade mortar or equivalent.(if required)
- Applying Filler board of - Dura Board HD 100 - Make

by Supreme in expansion joint area, etc, complete as per site conditions.(if required)

- Applying Highly elastomeric crack bridging waterproofing coating/Two Component Resin-Based, Thixotropic, Multipurpose Adhesive First Coat as a Primer Coat for Applying tape about the expansion joint and both sides surrounding area.
- Applying Imported Sealing tape / High Performance Expansion Joint Sealing Tape or Equivalent on primer coat about expansion joint and both side surround area.
- Applying highly elastomeric crack bridging water proofing coating/Two Component Resin-Based, Thixotropic, Multipurpose Adhesive First Coat, Second Coat on tape about only both side surround area, keeping expansion joint area without coating to make it expand.
- Applying highly elastomeric crack bridging waterproofing coating on Imported Sealing tape Third & Final Coat on tape about only both sides of the surrounding area, keeping the expansion joint area without coating to make it expand.

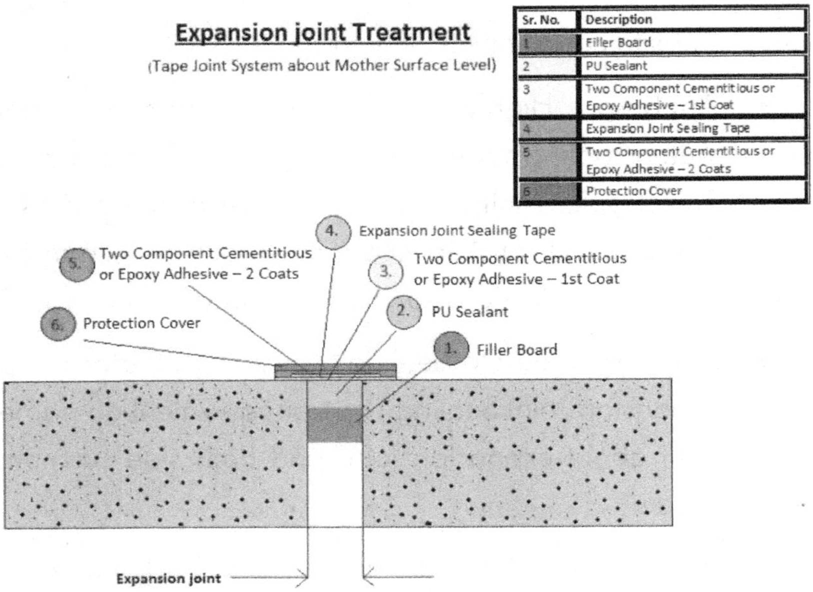

Fig. 5.5.2 Expansion Joint Treatment

Step-5 (Expansion Joint at Junction of Wall & Floor Area):

- Applying Coating Work for Expansion: (PU Sealant + Self Penetrating Primer + UV Stable Acrylic Heavy Duty Waterproof Coating + Net + UV Stable Acrylic Heavy Duty Waterproof Coating – 2 Coats):
- Applying PU Sealant Filling Work at Junction as per site requirement.
- Applying Self Penetrating Primer as a 1st Layer.
- Applying 1st coat of UV Stable Acrylic Heavy Duty Waterproof Coating as a 2nd layer.

- Applying Fibre Net/Non-Woven Geotextile Net on the 2nd layer as a 3rd layer.
- Applying 2nd coat of UV Stable Acrylic Heavy Duty Waterproof Coating as a 4th layer.
- Applying 3rd coat of UV Stable Acrylic Heavy Duty Waterproof Coating as a 5th layer.

Step – 6- Applying India Waterproofing (IWP)/IPS/ Screeding area as per site requirement

Applying Geotextile/Fibre Net on Floor + Applying IWP/ IPS/Screeding on Floor Area with Integral Waterproofing Compound.

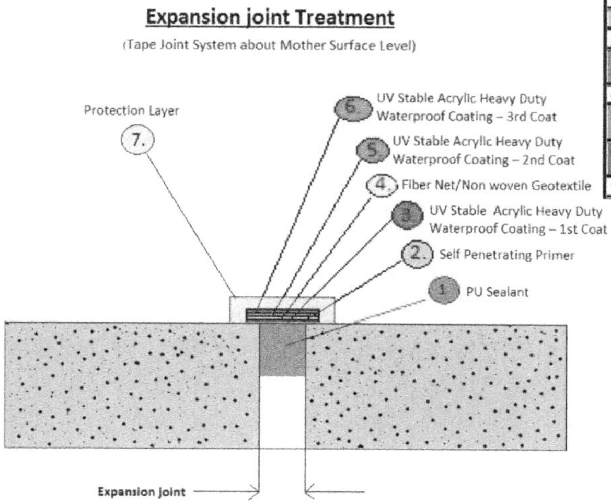

Sr. No.	Description
1	PU Sealant
2	Self Penetrating Primer
3	UV Stable Acrylic Heavy Duty Waterproof Coating – 1st Coat
4	Fiber Net/Non woven Geotextile
5	UV Stable Acrylic Heavy Duty Waterproof Coating – 2nd Coat
6	UV Stable Acrylic Heavy Duty Waterproof Coating – 3rd Coat
7	Protection Layer

Fig. 5.5.3 Expansion Joint Treatment at Floor & Wall after finished layer

Please scan the QR Code for further information:

"Alone we can do so little; together we can do so much."

- Helen Keller

5.6 Under Ground Water Tank (UGWT) Waterproofing

Underground water tanks are vulnerable to water ingress from external sources like surrounding soil as well as soluble salts can infiltrate the tank and cause reinforcement attack if not properly waterproofing is carried out. Our comprehensive waterproofing solutions can keep your structure free from any harmful chemical attack.

Specification: System of Waterproofing

Step-1: Borehole Packing Work

The Tank Area to be waterproofed must be cleaned first, as RCC & other debris are pending in the Wall & Junction area to be Waterproofed.

Epoxy injection for Strengthening & Plumbing done before waterproofing work

Surfaces to be waterproofed shall be sound, clean, and free of dust, paint, Curing agents, loose plaster debris, etc.

All final cleaning will be carried out with an electric blower & Water jet (if required).

Finally, all the surface areas shall be dampened thoroughly with clean water, as this is necessary prior to the application of any type of coating.

- Borehole Packing System - Borehole Packing Work with Non-Shrink Free Flow Grout & Imported Joint

Tape System + Plaster/PMM/Vata + Geotextile with Coating.

- Crack filling/PMM work as per Site Conditions.

<u>Step-2: Active Leakage Area – Injection Grouting Work (If required) – PU / Cementitious</u>

- Demolition or Removal of existing damaged/broken/ loose R.C.C area.
- Drilling 14mm diameter holes with electric drilling machines along the honey combed area and if required at floor wall joints slab and fixing Aluminium or G.I. NRV heavy nozzles of 10 mm diameter in place using quick setting compound.
- Filling Polymer Modified Mortar in Honey Combed R.C.C/Brick Masonry Area after finalising & fixing the Nozzles.
- Pressure grouting should be carried out by using [PU FOAM] Hand operated Centrifugal PU grouting machine through the nozzles, starting from one nozzle and stopping the operations when the grout starts coming out from the adjacent nozzle; this is done similarly in all the nozzles. [As per client's suggestion or actual site condition requirement]
- After Completion of PU Grouting work, Remove or Cut the Aluminium or G.I. NRV heavy nozzles & plug

the holes filled by quick setting compound or Non-Shrink Free Flow Cementitious Grout/ Repair Mortar /Polymer Modified Mortar (PMM).

- Providing & Applying Crystalline Waterproofing -2 Coats on only nearby the Nozzles surrounding area.
- Make curing or apply curing compound to that area. [If required]

OR

- Applying Cementitious Injection Grouting Work as per Site requirement. [if required]

Step 3: Imported Tape Joint Treatment

- Cleaning Work as per requirement.
- Crack filling/PMM / SBR Latex work as per Site Conditions.
- Imported Tape Joint System with Adhesive at All Junction Areas.

Step-4: Waterproofing Coating Work with Six Options

SBR Latex Acrylic – 3 Coats (Only basic waterproofing purpose)

OR

Two Component Flexible Crack Bridging Waterproof Coating - 2 Coat (Primer + Net + Two component - 2 Coats) – 4 Layer Systems

OR

Two Component Flexible Crack Bridging Waterproof Coating - 3 Coat (Primer + Net + Two components - 3 Coats) – 5 Layer Systems

OR

Crystalline Waterproofing – 3 Coat

OR

Bituminous Coating – 3 Coats + Sand Sprinkler

OR

Single Component PU Coating (Primer + 2 Coats)

Step – 5: Applying India Waterproofing (IWP)/IPS/ Screeding at Floor area & PMM/Plaster at Wall as per site requirement

Apply Geotextile/Fibre Net on the Floor + Apply IWP/IPS/ Screeding on the Floor Area & PMM/Plaster at the Wall with Integral Waterproofing Compound.

Step-6: CFTRI Approved /Food Grade Coating

CFTRI Approved coating – 3 Coats (Primer + 2 Coats) on Finish Plaster for making water food grade proof. (if required)

OR

Step-6: Tiles with Tile Mortar (Adhesive) with tile Grout

Tiles with Tile Mortar (Adhesive) with Tile Grout and filled with Epoxy or Cementitious Grout material.

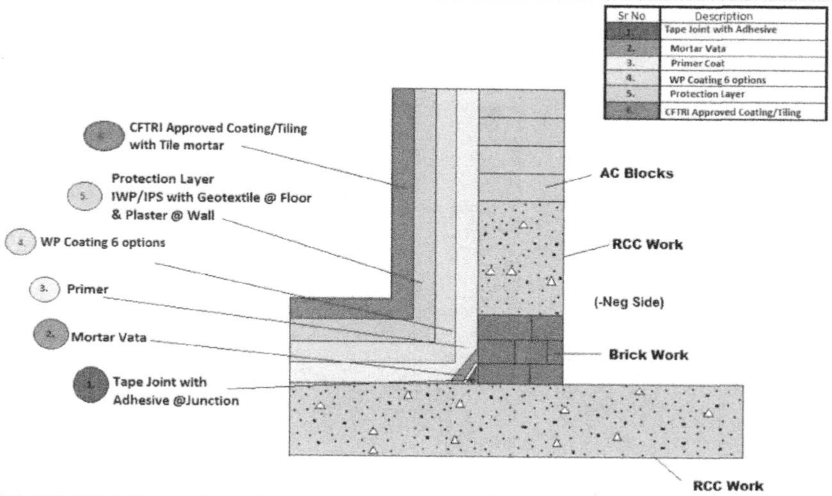

Fig. 5.6 UGWT (Underground Water Tank) Waterproofing

Please scan the QR Code for further information:

"Opportunities don't happen, you create them."

- Chris Grosser

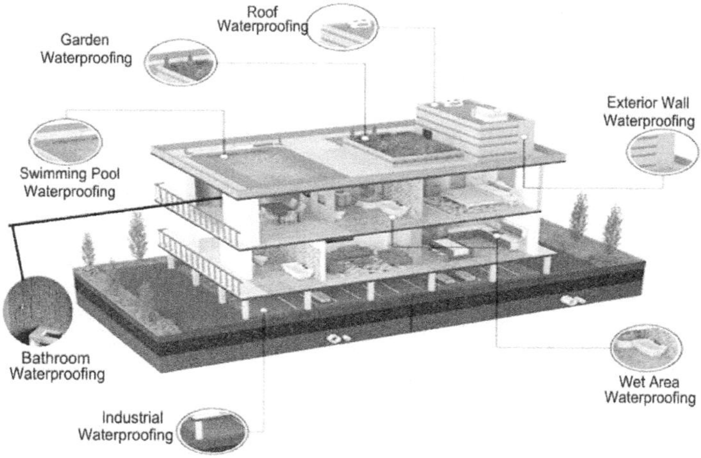

Chapter

6

Super Structure Waterproofing

6.1 Podium Area: -

The podium is the base of any building. It is divided into two types.

1. Landscape Podiums: - (Garden / Planters)
2. Hardscape Podiums: - (Driveway Area)

6.1.1 Landscape Waterproofing: -

To avoid any kind of leakage garden/planters and landscape are required waterproofing. This is because they are exposed to water in any season.

Now, if this is not done, then it will create serious risk to the building & it will be difficult to correct it at a later stage.

According to WHO, water leakage in buildings creates serious health-related problems like asthma, fungal infections, etc.

Specification: System of Waterproofing

Step-1: Cleaning Work & Crack filling

Surfaces to be waterproofed shall be sound, clean, free of dust, paint, Curing agents, loose plaster/R.C.C. debris etc.

All final cleaning will be carried out with an electric blower & Water jet (if possible).

Finally, all the surface areas shall be dampened thoroughly with clean water, as this is necessary prior to the application of any type of coating.

Applying Crack filling Work with Crack Filler & Polymer Modified Mortar (P.M.M.) @ Junction & Honeycombed Area with S.B.R.: Water: Cement (For detail Refer Chapter No: 7.2).

Step 2: Cold/Construction Joint Area Treatment

- Cleaning Work as per requirement.
- Applying Crack filling/P.M.M. **(Polymer Modified Mortar)** work as per Site Conditions.
- Applying PU Sealant as per site condition.
- Applying Sandwich coat of Coating with 150 GSM Fiber Mesh on Cracks/Cold/Construction Joints.
- Applying Imported Tape Joint System with Adhesive @ All Junction Area.

Step-3: Active Leakage Area – Injection Grouting Work (If required) – P.U. / Cementitious

- Go for the Naked Pond Test Before waterproofing work.
- If leakage is found, go for Injection Grouting from the Negative side.

P.U. Injection Grouting

- Demolition or Removal of existing damaged/broken/ loose R.C.C. area. [if required]
- Drilling 14mm diameter holes with electric drilling machines along the honeycombed area and, if required @ floor wall joints slab and fixing Aluminium or G.I. N.R.V. heavy nozzles of 10 mm diameter in place using quick setting compound Insta-stop/Water Plug/S.B. (L).
- Providing & Filling Polymer Modified Mortar in Honey Combed R.C.C./Brick Masonry Area after finalising & fixing the Nozzles. [if required]
- Pressure grouting should be carried out by using [PU FOAM] Hand operated Centrifugal PU grouting machine through the nozzles, starting from one nozzle and stopping the operations when the grout starts coming out from the adjacent nozzle; this is done similarly in all the nozzles. [As per client's suggestion or actual site condition requirement]
- Pressure grouting should be carried out by using [P.U. Resins] Hand operated/Centrifugal PU grouting machine through the nozzles, starting from one nozzle and stopping the operations when the grout starts coming out from the adjacent nozzle; this is done

similarly in all the nozzles. [As per client's suggestion or actual site condition requirement]

- After Completion of P.U. Grouting work, Remove or Cut the Aluminium or G.I. N.R.V. heavy nozzles & plug the holes filled by Insta-stop or Non-Shrink Free Flow Cementitious Grout/ Repair Mortar /Polymer Modified Mortar (P.M.M.).
- Providing & Applying Crystalline Waterproofing -2 Coats on only the Nozzles & surrounding area.
- Make curing or apply curing compound to that area. [If required]

OR

- Applying Cementitious Injection Grouting Work as per Site requirement (Use only as per site condition; try to go for P.U. Injection Grouting as 1st Option)

Cementitious Injection Grouting

- Demolition or Removal of existing damaged/broken/ loose R.C.C. area. [if required]
- Drilling 14mm diameter holes with electric drilling machines along the honeycomb area and, if required, @ floor wall joints slab and fixing P.V.C. nozzles of 10 mm diameter in place using quick setting compound Insta-stop/Water Plug/S.B. (L).

- Providing & Filling Polymer Modified Mortar in Honey Combed R.C.C./Brick Masonry Area after finalising & fixing the Nozzles. [if required]
- Pressure grouting should be carried out by using [Cement, Expanding Grout, S.B.R., Water] Hand operated/ Centrifugal PU grouting machine through the nozzles, starting from one nozzle and stopping the operations when the grout starts coming out from the adjacent nozzle; this is done similarly in all the nozzles. [As per client's suggestion or actual site condition requirement]
- After Completion of the Grouting work, Remove or Cut the P.V.C. nozzles & plug the holes filled by Insta-stop or Non-Shrink Free Flow Cementitious Grout/ Repair Mortar /Polymer Modified Mortar (P.M.M.).
- Providing & Applying Crystalline Coating -2 Coats on only nearby the Nozzles area.
- Make curing or apply curing compound to that area. [If required]

Step-4: Waterproofing Coating work with 6 Options

Laying APP Membrane (3mm to 4mm thick)

OR

Applying Bituminous Coating – 2 or 3 Coats with Fiber Mesh/Geotextile

OR

Applying Crystalline Coating – 2 or 3 Coats

OR

Applying Two Component – 3 Coats (Anti Root Coating)

OR

Applying PU Coating – 2 Coats with Primer

OR

Poly Urea Coating System – 2 Coats with Primer

- Making Corner fillets of 50 mm x 50 mm using polymer-modified mortar added with Multipurpose Polymer.
- All construction joints will be treated using polymer-modified mortar added with Multipurpose Polymer & Grouting wherever required using Cement slurry added with Grouting Admixture until refusal.
- All cracks will be treated using Non-shrink grout/ Epoxy mortar.
- The area to be waterproofed shall be ponded with water for a minimum of 24 hours, and maximum upto 72 hours, leakage/ dampness, if any, shall be marked. Nipples shall be fixed at

weak locations, wherever leakage persists, along construction joints, and with pressure grouting using neat cement slurry admixed with grouting additive.

- The surface shall then be primed with Epoxy Primer.
- Supplying and applying Solvent free Two component Hybrid Polyurea Polyurethane Liquid membrane with an elongation of >600% as per ASTM D 2370, tensile strength >10Mpa as per ASTM D 2370 and Shore A hardness of 85 as per ASTM D2240. Apply hybrid polyurea using a plural component, a high temperature / high-pressure proportionate unit fitted with a mechanically purge impingement mix spray gun. Minimum consumption shall be 2.0 kg/sqm to form a DFT of 1.5 to 2 mm. The membrane will be terminated a minimum of 300 mm above Finish Floor Level (FFL).

Step-5: Protection Layer

- Applying 120 GSM Non-Woven Geotextile on the Floor Area.
- Applying I.P.S./Screed/Brick Bat Coba or India Waterproofing (I.W.P.) as per Site Conditions.

Step-6: Separation Layer

- Applying Drain cell/Drain Board on Floor Area & a certain height of wall area, as per site situation.
- Applying Non-Woven Geotextile Net with Adhesive as a spot welding.

Sr No:	Description
1.	Primer
2.	Waterproofing Coating 5 Options
3.	IWP/IPS Srceed
4.	Drain Cell
5.	Non Woven Geotextile
6.	Soil/Garden

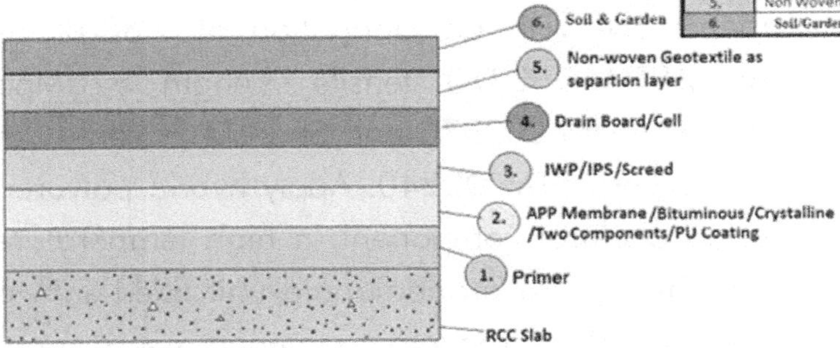

Fig. 6.1.1 Landscape-Garden Area Waterproofing

Please scan the Q.R. Code for further information:

"All our dreams can come true if we have the courage to pursue them."

- Walt Disney

6.1.2 Hardscape Waterproofing: -

If this is done properly & through the waterproofing expert, it will not have any problems for a number of years as well as you can protect your investment. You will also have stain protection. This is very much useful in rainy season.

Another reason of sealing your driveway is that, it will decrease the chances of cracks which happen due to water penetrating and later on expansion when it freezes.

Why Driveway / Hardscape waterproofing is needed?

- To minimise cracks.
- To prevent liquid from seeping through the concrete
- To prevent the eroding of the soil.
- To prevent concrete from cracking under its weight.
- To prevent unsightly surface cracks.
- To prevent material degradation.
- To prevent fissures that will create serious problems later on.
- To prevent the need for repairs.

Specification: System of Waterproofing

Step-1: Cleaning Work & Crack filling

Surfaces to be waterproofed shall be sound, clean, free of dust, paint, Curing agents, loose plaster debris etc.

All final cleaning will be carried out with an electric blower & Water jet (if possible).

Finally, all the surface areas shall be dampened thoroughly with clean water, as this is necessary prior to the application of any type of coating.

Applying Crack filling Work with Crack Filler & Polymer Modified Mortar (P.M.M.) @ Junction & Honeycombed Area with S.B.R.: Water: Cement (Refer Chapter No:7.2).

Step-2: Cold/Construction/Expansion Joint Area Treatment

- Cleaning Work as per requirement.
- Applying crack filling/PMM work as per Site Conditions.
- Applying PU Sealant as per site condition.
- Applying Sandwich coat of Coating with 150 GSM Fiber Mesh on Cracks/Cold/Construction Joints.
- Applying Imported Tape Joint System with Adhesive @ All Junction Area.

Step-3: Active Leakage Area – Injection Grouting Work (If required) – P.U. / Cementitious

- Go for the Naked Pond Test Before waterproofing work.
- If leakage is found, go for Injection Grouting from the Negative side.

P.U. Injection Grouting

- Demolition or Removal of existing damaged/broken/ loose R.C.C. area. [if required]
- Drilling 14mm diameter holes with electric drilling machines along the honeycombed area and, if required @ floor wall joints slab and fixing Aluminium or G.I. N.R.V. heavy nozzles of 10 mm diameter in place using quick setting compound Insta-stop/Water Plug/S.B. (L).
- Providing & Filling Polymer Modified Mortar in Honey Combed R.C.C./Brick Masonry Area after finalising & fixing the Nozzles. [if required]
- Pressure grouting should be carried out by using [PU FOAM] Hand operated Centrifugal PU grouting machine through the nozzles, starting from one nozzle and stopping the operations when the grout starts coming out from the adjacent nozzle; this is done similarly in all the nozzles. [As per client's suggestion or actual site condition requirement]
- Pressure grouting should be carried out by using [P.U. Resins] Hand operated/Centrifugal PU grouting machine through the nozzles, starting from one nozzle and stopping the operations when the grout starts coming out from the adjacent nozzle; this is done similarly in all the nozzles. [As per client's suggestion

or actual site condition requirement]

- After Completion of P.U. Grouting work, Remove or Cut the Aluminium or G.I. N.R.V. heavy nozzles & plug the holes filled by Insta-stop or Non-Shrink Free Flow Cementitious Grout/ Repair Mortar /Polymer Modified Mortar (P.M.M.).
- Providing & Applying Crystalline Waterproofing - 2 Coats on only the Nozzles & surrounding area.
- Make curing or apply curing compound to that area. [If required]

OR

- Applying Cementitious Injection Grouting Work as per Site requirement (Use only as per site condition; try to go for P.U. Injection Grouting as 1st Option)

Cementitious Injection Grouting

- Demolition or Removal of existing damaged/broken/ loose R.C.C. area. [if required]
- Drilling 14mm diameter holes with electric drilling machines along the honeycomb area and, if required, @ floor wall joints slab and fixing P.V.C. nozzles of 10 mm diameter in place using quick setting compound Insta-stop/Water Plug/S.B. (L).
- Providing & Filling Polymer Modified Mortar in Honey

Combed R.C.C./Brick Masonry Area after finalising & fixing the nozzles.[if required]

- Pressure grouting should be carried out by using [Cement, Expanding Grout, S.B.R., Water] Hand operated/ Centrifugal PU grouting machine through the nozzles, starting from one nozzle and stopping the operations when the grout starts coming out from the adjacent nozzle; this is done similarly in all the nozzles. [As per client's suggestion or actual site condition requirement]
- After Completion of the Grouting work, Remove or Cut the P.V.C. nozzles & plug the holes filled by Insta-stop or Non-Shrink Free Flow Cementitious Grout/ Repair Mortar /Polymer Modified Mortar (P.M.M.).
- Providing & Applying Crystalline Coating -2 Coats on only nearby the Nozzles area.
- Make curing or apply curing compound to that area. [If required]

Step-4: Waterproofing Coating work with 5-Options

Laying App Membrane (3 mm to 4 mm thick)

OR

Applying Bituminous Coating - 2 or 3 Coats with Fiber Mesh/ Geotextile

OR

Applying Crystalline Coating – 2 or 3 Coats

OR

Applying Two Component – 3 Coats

OR

Applying PU Coating – 2 Coats with Primer

Step-5: Protection Layer

- Applying 120 GSM Non-Woven Geotextile on the Floor Area.
- Applying I.P.S./Screed/Brick Bat Coba or India Waterproofing (I.W.P.) as per Site Conditions
- Then after pond test you can able to go for trimix, granite or tiling work as per the architect's requirement.

Sr. No.	Description
1	Primer Coat
2	APP Membrane / Crystalline/ Bituminous/ Two Components/ PU Coating
3	IWP/IPS/Screed with Non-woven Geotextile
4	Trimix/Granite/Tilling Work

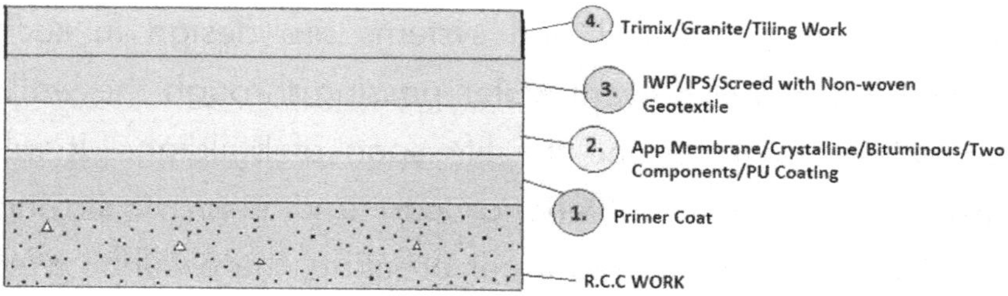

Fig. 6.1.2 Hardscape Area Waterproofing

Please scan the Q.R. Code for further information:

"Do one thing every day that scares you."

- Eleanor Roosevelt

6.2 Wet Area

Wet area waterproofing means waterproofing the bathroom, washing and other areas. As such, there is no need to explain why one should go for waterproofing of wet areas.

Waterproofing material and systems are design in such a way that it will prevent water passing through the walls as well as it will extend the life span of buildings. It will also reduce your maintenance cost and improve staying conditions. Without doing water proofing, there will be a lot of damage like paint failure, corrosion of metal structures, rotting of wood and huge repairing cost etc.

6.2.1 Toilet / Bathroom / Kitchen Wash

Specifications: System of Waterproofing

Step-1: Cleaning Work & Crack filling

Surfaces to be waterproofed shall be sound, clean, free of dust, paint, Curing agents, loose plaster debris etc.

All final cleaning will be carried out with an electric blower & Water jet (if possible).

Finally, all the surface areas shall be dampened thoroughly with clean water, as this is necessary prior to the application of any type of coating.

Applying Crack filling Work with Crack Filler & Polymer Modified Mortar (P.M.M.) @ Junction & Honeycombed Area with S.B.R.: Water: Cement (Refer Chapter:7.2).

Step 2: Cold/Construction Joint Area Treatment

- Cleaning Work as per requirement.
- Applying crack filling/P.M.M. work as per Site Conditions.
- Applying Imported Tape Joint System with Adhesive @ All Junction Area.

Step-3: Borehole Packing System

- P/A Borehole Packing Work with Non-Shrink Free Flow Grout & Imported on (junction area) Joint Tape System + Plaster/P.M.M./Vata + Geotextile with Coating. (with Sandwich coat)

Step-4: Active Leakage Area – Injection Grouting Work (If required) – P.U. / Cementitious

PU Injection Grouting

- Demolition or Removal of existing damaged/broken/ loose R.C.C. area. [if required]
- Drilling 14mm diameter holes with electric drilling machines along the honeycombed area and, if required @ floor wall joints slab and fixing Aluminium or G.I. N.R.V. heavy nozzles of 10 mm diameter in place using quick setting compound Insta-stop/Water Plug/S.B. (L).

- Providing & Filling Polymer Modified Mortar in Honey Combed R.C.C./Brick Masonry Area after finalising & fixing the Nozzles. [if required]
- Pressure grouting should be carried out by using [PU FOAM] Hand operated Centrifugal PU grouting machine through the nozzles, starting from one nozzle and stopping the operations when the grout starts coming out from the adjacent nozzle; this is done similarly in all the nozzles. [As per client's suggestion or actual site condition requirement]
- Pressure grouting should be carried out by using [P.U. Resins] Hand operated/Centrifugal PU grouting machine through the nozzles, starting from one nozzle and stopping the operations when the grout starts coming out from the adjacent nozzle; this is done similarly in all the nozzles. [As per client's suggestion or actual site condition requirement]
- After Completion of P.U. Grouting work, Remove or Cut the Aluminium or G.I. N.R.V. heavy nozzles & plug the holes filled by Insta-stop or Non-Shrink Free Flow Cementitious Grout/ Repair Mortar /Polymer Modified Mortar (P.M.M.).
- Providing & Applying Crystalline Waterproofing -2 Coats on only the Nozzles & surrounding area.
- Make curing or apply curing compound to that area. [If required]

OR

- Applying Cementitious Injection Grouting Work as per Site requirement (Use only as per site condition; try to go for P.U. Injection Grouting as 1st Option)

<u>Cementitious Injection Grouting</u>

- Demolition or Removal of existing damaged/broken/ loose R.C.C. area. [if required]
- Drilling 14mm diameter holes with electric drilling machines along the honeycomb area and, if required, @ floor wall joints slab and fixing P.V.C. nozzles of 10 mm diameter in place using quick setting compound Insta-stop/Water Plug/S.B. (L).
- Providing & Filling Polymer Modified Mortar in Honey Combed R.C.C./Brick Masonry Area after finalising & fixing the Nozzles. [if required]
- Pressure grouting should be carried out by using [Cement, Expanding Grout, S.B.R., Water] Hand operated/ Centrifugal PU grouting machine through the nozzles, starting from one nozzle and stopping the operations when the grout starts coming out from the adjacent nozzle; this is done similarly in all the nozzles. [As per client's suggestion or actual site condition requirement]
- After Completion of the Grouting work, Remove or Cut the P.V.C. nozzles & plug the holes filled by Insta-

stop or Non-Shrink Free Flow Cementitious Grout/ Repair Mortar /Polymer Modified Mortar (P.M.M.).

- Providing & Applying Crystalline Coating -2 Coats on only nearby the Nozzles area.
- Make curing or apply curing compound to that area. [If required]

Step-4: Waterproofing Coating work with 5-Options

Applying Two Component Semi Flexible Coating – 2 Coats

OR

Applying Two Component Highly Flexible Coating– 2 Coats with Fiber Mesh

OR

Applying Rubber base single Component Highly Flexible coating– 2 Coats

OR

Applying Liquid Applied Membrane – 2 Coats

OR

Applying Basic Coating of S.B.R.: Water: Cement (1: 1: 2) – 2 Coats

Step-5: Protection Layer

- Applying Fiber Net/Non-Woven on Floor Area.

- Applying I.P.S./ Screed/ Brick Bat Coba as per Site Conditions.

Sr. No.	Description
1	Borehole Packing with Tape & Cementitious Grout
2	Angle/Fillet with Tape/Fiber Net/Non-woven Geotextile
3	Two Component Semi flexible/Highly Flexible/Rubber Based/LAM Bituminous Coating/SBR:Water Cement -1:1:2 -2 Coats
4	Protection Layer

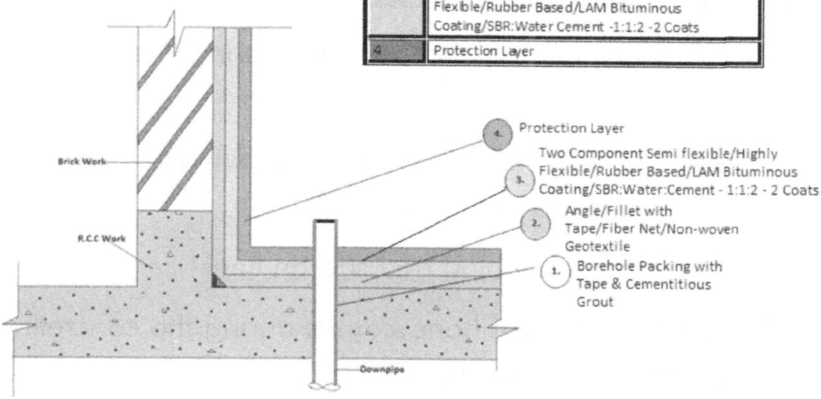

Fig. 6.2.1 Toilet / Bathroom / Kitchen Wash Area Waterproofing

Please scan the Q.R. Code for further information:

"Impossible is just an opinion."

\- Paulo Coelho

6.2.2 Swimming Pool/ Splash Pool/ Overhead Water Tank (OHWT)

Why Waterproofing of Swimming Pool/Splash Pool/ OHWT Required?

Factors that can lead to leakages are:

1. Shrinkage cracks that appear during the concrete curing process
2. Lack of waterproofing barriers between the pool water and concrete wall
3. Failures in plumbing and suction lines
4. The formation of honeycombs during the concreting process which weakens the structure
5. Construction joints
6. Pipe Penetrations

Specifications: System of Waterproofing

Step-1: Cleaning Work & Crack filling

Surfaces to be waterproofed shall be sound, clean, free of dust, paint, Curing agents, loose plaster debris etc.

All final cleaning will be carried out with an electric blower & Water jet (if possible).

Finally, all the surface areas shall be dampened thoroughly with clean water, as this is necessary prior to the application of any type of coating.

Applying Crack filling Work with Crack Filler & Polymer Modified Mortar (P.M.M.) @ Junction & Honeycombed Area with S.B.R.: Water: Cement (Refer Chapter 7.2).

Step-2: Cold/Construction/Joint Area Treatment

- Cleaning Work as per requirement.
- Applying crack filling/PMM work as per Site Conditions.
- Applying Imported Tape Joint System with Adhesive @ All Junction Area.

Step-3: Borehole Packing System

- Borehole Packing Work with Non-Shrink Free Flow Grout & Imported Tape Joint System/Bituminous Tape/Geotextile on (junction area - Pipe & Floor/ Wall Area) with required adhesive of sandwich coat + Plaster/P.M.M./Screed/Vata (if req.)
- Borehole filling in first stage (Up to half depth of slab approx.) filling by non-shrink free flow cementitious grout.
- Borehole filling is in the second and final stage (about half the depth of the slab to the floor).
- Finishing of the borehole from top and bottom, filled by Non-shrink free flow cementitious grout. (as required)

- After finishing, apply with Imported Tape Joint System/Bituminous Tape/Geotextile on (junction area- Pipe & Floor/Wall Area) with required adhesive of a sandwich coat.
- P.M.M./Plaster/Screed/Vata @ wall & floor (if req.).

Step-4: Active Leakage Area – Injection Grouting Work (If required) – P.U. / Cementitious

PU Injection Grouting

- Demolition or Removal of existing damaged/broken/ loose R.C.C. area. [if required]
- Drilling 14mm diameter holes with electric drilling machines along the honeycombed area and, if required @ floor wall joints slab and fixing Aluminium or G.I. N.R.V. heavy nozzles of 10 mm diameter in place using quick setting compound Insta-stop/Water Plug/S.B. (L).
- Providing & Filling Polymer Modified Mortar in Honey Combed R.C.C./Brick Masonry Area after finalising & fixing the Nozzles. [if required]
- Pressure grouting should be carried out by using [PU FOAM] Hand operated Centrifugal PU grouting machine through the nozzles, starting from one nozzle and stopping the operations when the grout starts coming out from the adjacent nozzle; this is done

similarly in all the nozzles. [As per client's suggestion or actual site condition requirement]

- Pressure grouting should be carried out by using [P.U. Resins] Hand operated/Centrifugal PU grouting machine through the nozzles, starting from one nozzle and stopping the operations when the grout starts coming out from the adjacent nozzle; this is done similarly in all the nozzles. [As per client's suggestion or actual site condition requirement]
- After Completion of P.U. Grouting work, Remove or Cut the Aluminium or G.I. N.R.V. heavy nozzles & plug the holes filled by Insta-stop or Non-Shrink Free Flow Cementitious Grout/ Repair Mortar /Polymer Modified Mortar (P.M.M.).
- Providing & Applying Crystalline Waterproofing - 2 Coats on only the Nozzles & surrounding area.
- Make curing or apply curing compound to that area. [If required]

OR

Cementitious Injection Grouting

- Demolition or Removal of existing damaged/broken/ loose R.C.C. area. [if required]
- Drilling 14mm diameter holes with electric drilling machines along the honeycomb area and, if required,

@ floor wall joints slab and fixing P.V.C. nozzles of 10 mm diameter in place using quick setting compound Insta-stop/Water Plug/S.B. (L).

- Providing & Filling Polymer Modified Mortar in Honey Combed R.C.C./Brick Masonry Area after finalising & fixing the Nozzles. [if required]
- Pressure grouting should be carried out by using [Cement, Expanding Grout, S.B.R., Water] Hand operated/ Centrifugal PU grouting machine through the nozzles, starting from one nozzle and stopping the operations when the grout starts coming out from the adjacent nozzle; this is done similarly in all the nozzles. [As per client's suggestion or actual site condition requirement]
- After Completion of the Grouting work, Remove or Cut the P.V.C. nozzles & plug the holes filled by Insta-stop or Non-Shrink Free Flow Cementitious Grout/ Repair Mortar /Polymer Modified Mortar (P.M.M.).
- Providing & Applying Crystalline Coating -2 Coats on only nearby the Nozzles area.
- Make curing or apply curing compound to that area. [If required]

Step-4: Waterproofing Coating work with 5-Options

Laying APP Membrane (3mm to 4mm thick)

OR

Applying Two Component Highly Flexible Coating– 3 Coats

OR

Applying crystalline coating – 3 Coats

OR

Applying Liquid Applied Membrane – 2 Coats

OR

Applying PU Coating – 2 Coats with Primer

Step-5: Protection Layer

- Applying Fiber Net/Non-Woven on Floor Area.
- Applying I.P.S./Screed/Brick Bat Coba as per Site Conditions.

After protection layer go for Pond Test for 7 days & check it out from all possible sides whether all is ok or not, if ok go for further finishing activity as per site condition & if not consult waterproofing consultant & as per his advice go for repairing of it & make it clear in all way.

Sr. No.	Description
1	Primer Coat
2	App Membrane/ Crystalline/ Bituminous/ Two Components/ PU Coating – 2 Coats
3	Protection Layer IPS/Screed @Floor & Plaster @ Wall
4	Glass Mosaic/ Tile with Tile Mortar & Epoxy Tile Grout

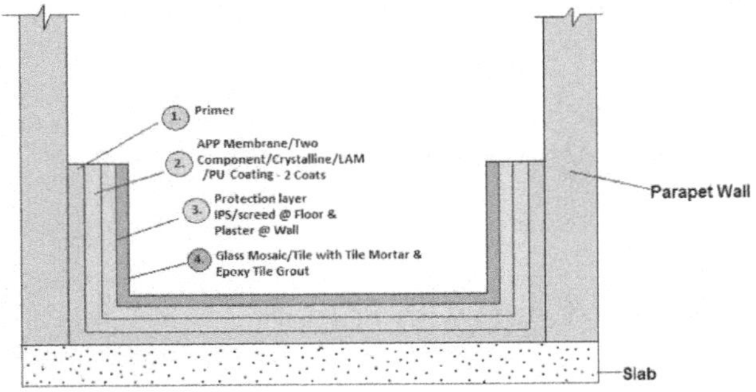

Fig. 6.2.2 Swimming Pool Waterproofing

Please scan the Q.R. Code for further information:

"If you believe it'll work out, you'll see opportunities. If you don't believe it'll work out, you'll see obstacles."

- Wayne Dyer.

6.3 Wall (Internal/External) Area

6.3.1 Rising Dampness/ Capillary Action

Dampness in wall occurs when moisture from the ground went up through the walls. This moisture contains white salt that also went up through walls. In other words, ground water went up via bricks /blocks /stones etc. Because of this, along with wall, other building material like plaster, timber, skirting etc. are also badly affected & damage occurs.

Why does this happen?

Because of failure of Damp proof course, it affects internal as well as external walls. It is therefore very much essential to have a Professional survey through a qualified waterproofing expert / consultant. He will carry out rising damp survey of your home / office including analysis of salts.

Please scan the Q.R. Code for further information:

Specifications: System of Waterproofing

[Step: 1] Loose /Existing old plaster or Substrate must be removed / demolished first with Electric Grinder up to 300 mm on all sides up level & side level of dampened area to be visualised.

{Step: 1 = 1 day activity} For Large Area

[Step: 2] Loose /Existing old plaster/P.O.P. Punning or any other surface breaking & removing up to original mother surface (Bricks/Stones/Block).

{Step: 2 = 2-day activity}

[Step: 3] Applying the bore holes shall be made by electric drill machine @ 12 to 14 diameter drill beat into wall up to certain depth @ ¾ of wall thickness i.e. for 14″ wall thickness depth @ 10″ in depth with the hole, Distance shall be kept up 4″ to 6″ c/c between each hole as per rising dampness of wall.

{Step: 3 = 1/2-day activity}

[Step: 4] Holes shall be cleaned by Blower/Vacuum cleaner, all Dust and loose particles shall be removed from holes.

{Step: 4 = 1/2-day activity}

[Step: 5] Applying Brick/ Block Masonry on all Joints with Extremely Setting Cement Powder or High-Performance, Self Curing Polymer-Modified Repair Mortar.

{Step: 5 = 1/2-day activity}

[Step: 6] Further, Holes shall be cleaned by Blower, all Dust and loose particles shall be removed from holes (if required).

{Step: 6 = 1/2-day activity}

[Step: 7] Applying Salt-Stopper (Solvent-free polymerised silicate liquid to bind salts in masonry and concrete) as a first coat by brushing as a net coat on Masonry/Mother surface wall, and allow to dry for 4 to 6 hrs. Then, apply the same as a second coat by brushing it with a net coat on the brick – Masonry wall and allow drying it for 4 to 6 hrs. Finally, apply the same as a third & final coat by brushing as a net coat on Brick – Masonry wall, and allow drying it for 4 to 6 hrs for further process of the system.

Salt Stoppers: 3 coats of the same material [with 4 to 6 hours of interval]

{Step: 7 = 2-day activity}

[Step: 8] Applying Two Component solvent-free or Single Component Crystalline Waterproofing (High Performance, Flexible, Self Curing, Cement–Based Crystalline System) as a first coat by brush as a primer coat on the Masonry/ Mother Surface wall, allow drying it for 6 to 8 hrs. Then apply the same as a second coat by brush as per Technical Data Sheet (T.D.S.) on Brick – Masonry wall, allow drying it for 6 to 8 hrs. Finally apply the third & final coat by brush on Brick – Masonry wall, allow drying it for 6 to 8 hrs for further process of the system.

Two Component solvent-free/ Single Component Crystalline Waterproofing: 3 coats of material [with 6 to 8 hours of interval]

{Step: 8 = 2-day activity}

[Step: 9] Injecting Impregnating Cream for Injection into Masonry or Concrete Walls against Capillary Rising Humidity

System for Rising Dampness and efflorescence into your wall sectional areas at skirting level by borehole method. The material shall be injected through suitable pump device into throughout depth of the Hole, until it's come out from the hole, keep the hole for same condition for 24 to 36 hrs for first time.

Inject the material after 24 to 36 hrs for the Second time until it comes out from the hole, and keep the hole in the same condition for 24 to 36 hrs for the second time.

Inject the material after 24 to 36 hrs for the third & final time until it comes out from the hole; plug the hole with.

{Step: 9 = 4-day activity}

[Step: 10] Applying Plaster in Cement –Sand Mortar (1: 4) with Integral Waterproofing Compound as a first coat as a Rough Coat Plaster.

{Step: 10 = 1 day activity}

[Step: 11] Applying Plaster in Cement –Sand Mortar (1: 4) with Integral Waterproofing Compound as a Second & Final coat as a Finish/Rough Plaster with 5 to 15 mm in depth for finishing finalisation, as per site requirements.

{Step: 11 = 1 day activity}

If these 11 steps are applied (small/big) for approximately 17 days with perfectly as per guided activity at any area then it will form an artificial D.P.C. (Damp Proof Course) that will make wall full proof from leakage/dampness/seepage.

Sr. No.	Description
1	Cement Base Joint Filler
2	Salt Stopper - 3 Coats
3	Crystalline Coating -3 Coats
4	Impregnating Cream for Injection
5	Plaster on Wall – 2 times

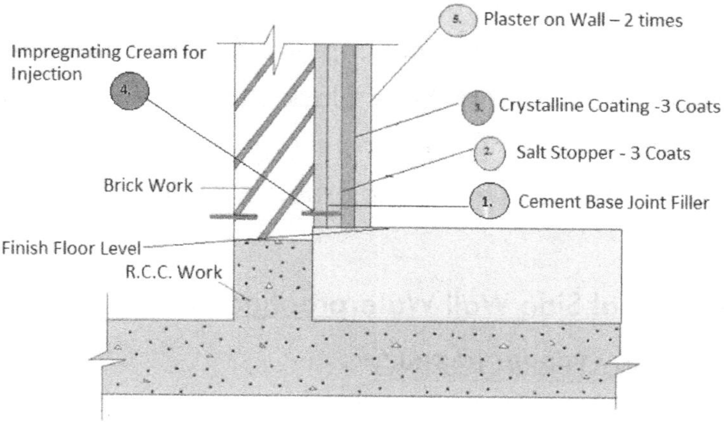

Fig. 6.3.1 Rising Dampness Treatment

Please scan the Q.R. Code for further information:

"There are only two ways to live your life. One is as though nothing is a miracle. The other is as though everything is a miracle."

- Albert Einstein.

6.3.2 Wall Waterproofing (Internal/External) Dampness

Structural dampness is the reason of unwanted moisture in the building. Further, the dampness also comes through climate change and rain penetration. Secondly, walls are becoming damp when internal symptoms become apparent. This symptom includes patches black mould growth, salt containment & damping the plaster. However, it is to be noted that the source of the damp problem could be external or internal.

6.3.2.1 Internal Side Wall Waterproofing:

SYSTEM OF WATERPROOFING

(Damp-Proof Coating for Internal Walls)

Loose / Existing old putty & colour work must be removed/ demolished First by Wire Brush / Sand Paper or with a Cup Cone Grinder.

Surfaces to be waterproofed shall be sound, clean, and free of dust, paint, Curing agents, loose plaster debris, etc. All final cleaning will be carried out with water & by electric blower (If required)

Applying (As per requirement) Gap Filling putty of Two Component Epoxy resin (As per T.D.S. Mixing Ratio as a putty: 1:1:2 by Vol (Base: Hardener: 2 O.P.C./Cement). Allow it to Dry for 24 Hours.

Applying the First Coat as a Primer Coat of Two Component Epoxy resin (Mixing Ratio as a Primer: 1:1:1 by Vol (Base: Hardener: Water).

After 6 to 8 Hours of 1st Coat, Apply a Second Coat of Two Component Epoxy resin (Mixing Ratio as a Coating: 1:1 by Vol (Base: Hardener).

After 6 to 8 Hours of 2nd Coat, Apply the Third & Final Coat of Two Component Epoxy resin [Mixing Ratio as a Coating: 1:1 by Vol (Base:Hardener)]

Then, allow it to dry for 3 to 7 days. (As per T.D.S.)

If you want to check it for dampness, then go through it.

After that, go for Colour/Putty for any kind of finishing work. This is CFTRI-approved Food Grade Certified, so there is nothing to worry about or any kind of contamination.

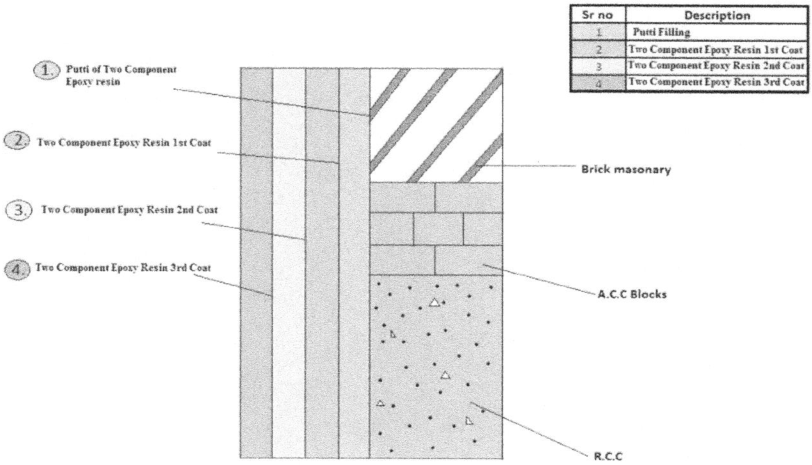

Fig. 6.3.2.1 Internal side walls Waterproofing Work

Please scan the Q.R. Code for further information:

"You cannot control the behaviour of others, but you can always choose how you respond to it."

- Roy T. Bennett

6.3.2.2 External Side Wall Waterproofing:

Why External Side Wall Waterproofing required in all kinds of buildings/projects/structures?

Due to weather & speed in today's construction, for prevents water ingress, maintain external looks of buildings, for increase durability, also increase structural integrity external side wall waterproofing is very very essential & can not neglect in any case.

Specifications: System of Waterproofing

Step-1: Cleaning Work & Crack filling

Surfaces to be waterproofed shall be sound, clean, free of dust, paint, Curing agents, loose plaster debris etc.

All final cleaning will be carried out with an electric blower & Water jet (if required).

Finally, all the surface areas shall be dampened thoroughly with clean water, as this is necessary prior to the application of any type of coating.

Applying Crack filling Work with Crack Filler & Polymer Modified Mortar (P.M.M.) @ Junction & Honeycombed Area with S.B.R.: Water: Cement or by Sandwich coating of S.B.R. or by P.U. Sealant as per site condition.

Step 2: Waterproofing Coating work

Applying Acrylic Based U.V. Stable Elastomeric Coating –3 Coats

Then, go for the actual external colour as per requirement before the transparent coating.

Step-3: Transparent Coating

Applying Transparent Coating –2 Coat (Wet on Wet)

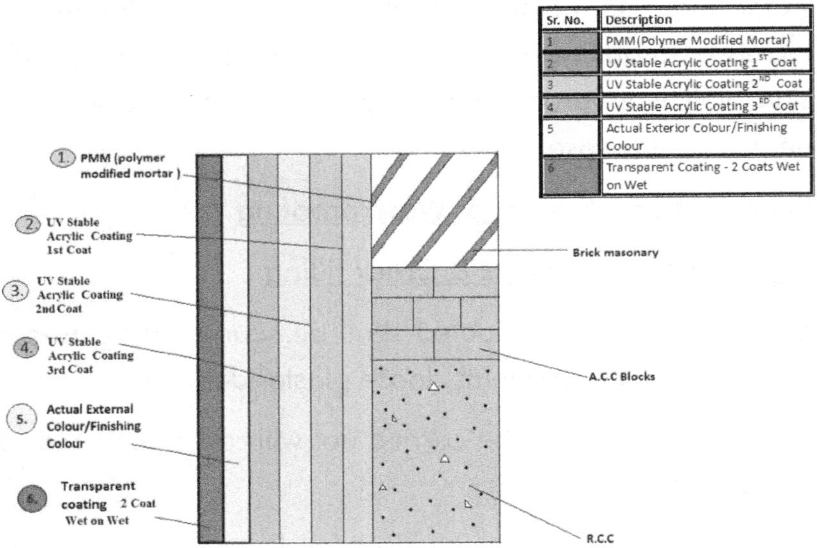

Sr. No.	Description
1	PMM(Polymer Modified Mortar)
2	UV Stable Acrylic Coating 1ST Coat
3	UV Stable Acrylic Coating 2ND Coat
4	UV Stable Acrylic Coating 3RD Coat
5	Actual Exterior Colour/Finishing Colour
6	Transparent Coating - 2 Coats Wet on Wet

Fig. 6.3.2.2 External wall Waterproofing

Please scan the Q.R. Code for further information:

"Be the change that you wish to see in the world."

- Mahatma Gandhi

6.3.3 Water Repellent Coating

For protecting building/structure/infra project from water ingress, algae, fungus, external outlook etc. water repellent coating is needed.

If this is done, the water does not stick to the surface, and it runs off from the water-repellent applied surface.

This type of coating is applied on

1. Exposed R.C.C. surface
2. Exposed Brick Masonry surface
3. Any kind of column / textured / stone surface

Specifications: System of Waterproofing

Surfaces to be waterproofed shall be sound, clean, free of dust, paint, Curing agents, loose plaster debris etc.

All final cleaning will be carried out with an electric blower & Water jet (if required).

Finally, all the surface areas shall be dampened thoroughly with clean water, as this is necessary prior to the application of any type of coating.

- Applying Crack filling/P.M.M. work as per site conditions and repair (if required) provide finishing then do water repellent coating.
- Applying Water base Repellent Coating – 2 Coats (Wet on Wet)

OR

- Applying Solvent base solvent-repellent coating – 2 Coats (wet on wet)

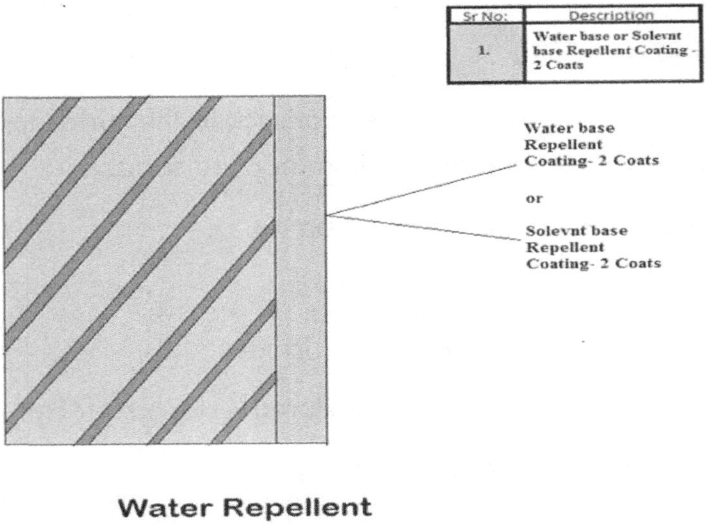

Fig. 6.3.3 External Wall Water Repellent Coating

Please scan the Q.R. Code for further information:

"Life isn't about finding yourself,
Life is about creating yourself."

- George Bernard Shaw

6.3.4 Dry Wall Waterproofing

Dry wall waterproofing means making the dry wall free from moisture & water damage. This is very useful in the areas of high humidity or straight exposure to water such as Bathrooms, Kitchens & laundry rooms.

This waterproofing is most important because in absence of it, there may be a problems of Mildew Growth (create health issue & damage to dry wall), Structural damage (weakening of dry wall & sometimes collapsing of dry wall), Aesthetic Damage (stains, bubbling & peeling of paint etc), Insulation compromise (affecting inside insulation, decrease in effectiveness leads to higher energy cost) etc.

Now, how to water proof Dry wall & which material to be used?

There are six techniques of doing waterproofing of Dry wall. They are as under.

(1) Moisture proof Dry wall (Green Board):

In this type, dry wall will have green coating which gives more resistance than normal dry wall. It is not actually waterproofing but it is useful where exposure to moisture is occasional.

(2) Cement Board:

It is also known as backer board & it is highly water resistant and hence it used as a base for tile in showers and tube surrounds.

3) Waterproof Membranes:

This is a waterproof barrier applied over the surface of dry wall. Hence, if creates waterproof barrier, This may be in the form of sheets or liquid applied coatings, Popular Products in this field is Red Gard & schluter - KERDI.

4) Water Resistant Coating & Paints:

This technique will provide additional layer of protection against moisture. This is designed to repel water and prevent it from penetrating the dry wall.

5) Sealing Joints & Seams:

Proper Sealing of joints & seams between dry wall panels is very much important for preventing water seepage. This can be done using waterproof tape & joint compound.

6) Proper Ventilation:

Proper Ventilation in areas prone to moisture, will help in reducing the risk of water damage to dry wall, This includes using exhaust fans & bathrooms & kitchens.

By using these six techniques & materials, you can enhance the moisture resistant of dry wall very significantly and you can extend its life span in areas exposed to water & humidity.

Please scan the Q.R. Code for further information:

"Nothing is impossible. The word itself says 'I'm possible'!"

- Audrey Hepburn.

6.4 Terrace/ Chhajja/ Balcony/ OTS/ Industrial Terrace

6.4.1 Terrace Waterproofing

Why Is Roof Waterproofing Important?

A roof or the terrace is the top covering of your home/ Structure, which protects you against rain and extreme weather conditions during summer and winter. An envelope that protects your home's interior and exterior, which is the most exposed part of your house. Normally, we expect the roof of our homes to be watertight and strong enough to withstand all the challenges nature throws in its way. But in reality, because of the constant exposure to extreme weather, roofs gradually get damaged. If we don't prevent this damage, the entire building could suffer from water leakage, seepage and cracks.

Durability

Proper waterproofing solutions last a long time, so if dampness and cracks appear on your roof and walls every year, even after waterproofing, your approach might need to be changed. With this in mind, it is important to go for professional waterproofing solutions and products that can provide elongation and last for years to come with the help of an Expert Waterproof Consultant.

WATERPROOFING FOR ROOFS, A MUST!

In terms of deterioration, water-related damage to a structure can be compared to fire. Like fire, water seepage through the roof first damages a home cosmetically and then eventually structurally.

Here are reasons to help you decide to waterproof your terrace:

Increases the lifespan of the roof

Prevents damage to structure

Maintains visual appeal of the home

Safeguard against health issues

Increasing Commercial values of your home/office

Specifications: System of Waterproofing

Step: 1-Debris Breaking & Removal

For Old Existing Terrace:

Existing China Mosaic and India Waterproofing and staking debris @ particular Points and Removal of it.

For the new terrace, clean it as much as the newly cast today.

Step-2: Waterproofing Tape Joint System @ All Junctions

(Micro Cleaning + Crack filling & PMM + Sealing Tape with Adhesive)

The Terrace Area to be waterproofed must be cleaned as R.C.C. & other debris are pending in the Wall & Junction area to be waterproofed.

Surfaces to be waterproofed shall be sound, clean, free of dust, paint, Curing agents, loose plaster debris etc.

All final cleaning will be carried out with an electric blower & Water jet (if required).

Finally, all the surface areas shall be dampened thoroughly with clean water, as this is necessary prior to the application of any type of coating.

- Cracks shall be cut down by mechanically in "V" shape, cleaned & filled up with a paste of Polymer Modified Mortar or Crack- X Powder/by P.U. Sealant/ Acrylic Sealant as per Site Conditions/requirement.
- Applying Junction of Floor & Wall Area Treatment with Sealing Tape with Two Component Acrylic based polymer modified Sandwich Coating.

<u>Step-3: Waterproofing Coating work of Terrace Area (For newly constructed Terraces)</u>

(Cleaning + Mortar Vata + Primer Coat + Fiber Net + Two Component Coating - 2 Coats = 4 Layers System)

- Applying Polymer Modified Mortar (P.M.M.) @ Junction & Honeycombed Area with S.B.R.: Water: Cement.
- Applying Mortar Vata

- Applying 1st Coat of S.B.R./Latex as a Primer Coat as a 1st layer.
- Applying Fiber Net on Primer Coat as a 2nd layer.
- Applying 1st Coat of Two Component Acrylic based polymer modified Coating as a 3rd layer.
- Applying 2nd Coat of Two-Component Acrylic-based polymer modified Coating as a 4th layer after 4 to 6 hours of 2nd Coat. (Apply any from the 5 Options)
- Always go for 300 mm high on the wall area for coating.
- Lay 120 GSM to 150 GSM Non-Woven Geotextile Fabric on the Floor Area.
- After that, lay the Thermal insulation board (25 to 50 mm thick as per the site conditions).
- At last, provide another layer of 120 GSM to 150 GSM Non-Woven Geotextile Fabric on it.

Waterproofing Coating Work with 5-Options

Laying APP Membrane (3mm to 4mm thick)

OR

Applying Two Component Acrylic Based Coating– 3 Coats

OR

Laying EPDM Membrane

OR

Applying Pure PU Coating– 2 Coats

OR

Applying Fast Setting PU Coating – 2 Coats with Primer

Step-4: Borehole Packing System:

- Borehole Packing Work with Non-Shrink Free Flow Grout & Imported Tape Joint System/ Bituminous Tape/ Geotextile on (Junction Area- Pipe & Floor/ Wall Area) with required adhesive of sandwich coat + Plaster/P.M.M./Screed/Vata (if req.)
- Borehole filling in first stage (Up to half depth of slab approx.) filling by non-shrink free flow cementitious grout.
- Borehole filling is in the second and final stage (about half the depth of the slab to the floor).
- Finishing of the borehole from top and bottom, filled by Non-shrink free flow cementitious grout. (as required)
- After finishing apply with Imported Tape Joint System/ Bituminous Tape/Geotextile on (junction area - Pipe & Floor/Wall Area) with required adhesive of sandwich coat.
- P.M.M./Plaster/Screed/Vata @ wall & floor (if req.).

Step-5: All Area India Waterproofing (Brick Bats Coba Concrete) IWP/IPS/Plaster:

- Applying Fiber Net/Geotextile after Coating Work & Before IWP/IPS/Screed.
- Applying India Water Proofing (I.W.P.)/I.P.S. / Screeding on floor area, applying P.M.M. (Polymer Modified Mortar) Or Plaster on wall area with Integral Water Proofing Compound.

Step-6: Applying for China Mosaic Work / Tilling Work

Applying China Mosaic Work with tools & tackles, etc., complete as per site conditions.

OR

Tile with spacers 3 to 5 mm with tile adhesive as per site conditions.

➢ For Existing Terrace Waterproofing Coating Work

- First of all, Hammer Test on entire Terrace & if found Hollow sound more than 25 % of it, you can follow steps as mentioned above.
- Do Waterproofing work as mentioned above.

OR

Step-3: Waterproofing Coating work of Terrace Area (For Existing Terraces)

- If only minor cracks are visible, then an acrylic-based coating of 3 coats shall be applied.

Sr No:	Description
1a.	Tape Joint Work
1b.	Cement mortar Vata
2	Primer
3	APP Membrane/Two Component Coating/EPDM Coating/PU Coating
4	Fiber Mesh
5	APP Membrane/Two Component Coating/EPDM Coating/PU Coating
6	After laying Thermal Insulation Board & Non-Woven Geotextile IWP(India waterproofing)/IPS
7	China Mosaic/Tiling Work

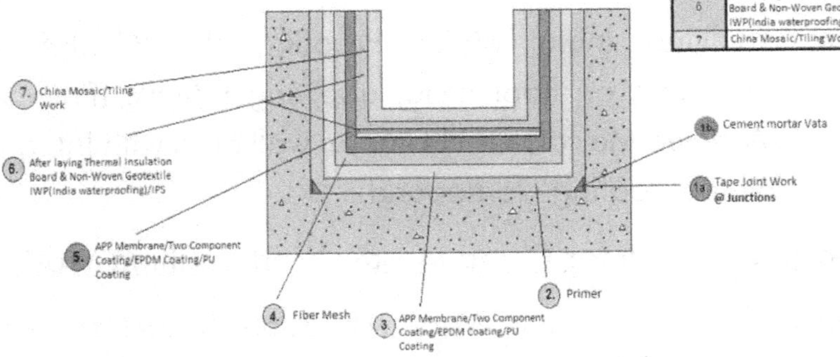

Fig. 6.4.1 Terrace Area Waterproofing Work

Please scan the Q.R. Code for further information:

"Life becomes easier and more beautiful when we can see the good in other people."

- Roy T. Bennett.

6.4.2 Chhajja Waterproofing

The word commonly known today as Chhajja is the projected element made in the building as a part of the design elevation given by the Architect for aesthetic purposes or part of structural design.

Waterproofing of Chhajja is crucial because it has to deal with harsh sun in summer & cold wind in winter, which leads to thermal expansion in the building & cracks can be formed in the junctions of Chhajja.

Chhajja are also filled with debris and close packed in some cases which can be serious issue if the rainwater gets collected in to this & cause leakage, so to avoid such leakage one should do waterproofing of Chhajja with extreme care & for filled Chhajja the provision of leakage pipe can be implement in it.

Specifications: System of Waterproofing

Step-1: Cleaning Work & Crack filling/ Borehole Packing Work

Surfaces to be waterproofed shall be sound, clean, free of dust, paint, Curing agents, loose plaster debris etc.

All final cleaning will be carried out with an electric blower & Water jet (if required).

Finally, all the surface areas shall be dampened thoroughly

with clean water, as this is necessary prior to the application of any type of coating.

Borehole Packing System: -

- Borehole Packing Work with Non-Shrink Free Flow Grout & Imported Tape Joint System/ Bituminous Tape/ Geotextile on (Junction Area- Pipe & Floor/ Wall Area) with required adhesive of sandwich coat + Plaster/P.M.M./Screed/Vata (if req.)
- Borehole filling in first stage (Up to half depth of slab approx.) filling by non-shrink free flow cementitious grout.
- Borehole filling in the second and final stage (about half the depth of the slab to the floor).
- Finishing of the borehole from top and bottom, filled by Non-shrink free flow cementitious grout. (as required)
- After finishing apply with Imported Tape Joint System/ Bituminous Tape/Geotextile on (Junction Area- Pipe & Floor/Wall Area) with required adhesive of sandwich coat.
- P.M.M./Plaster/Screed/Vata @ wall & floor (if req.).

Step-2: Imported Tape Joint/Fiber Net Treatment @ Floor & Wall Junctions

- Cleaning Work as per requirement.
- Applying crack filling/PMM work as per Site Conditions.
- Applying Fiber net of 2 to 4 inch-wide sandwich coat or Imported Tape Joint System with Adhesive @ All Junction Area.

Step 3: Waterproofing Coating work

- Applying Cement Mortar Vata.
- Applying Primer Coat.
- Applying Fiber Net on Primer Coat.
- Applying Waterproofing Coating

Step 3: Waterproofing Coating with 2 options

- For New Construction

Applying Two Component Based Coating – 2 to 3 Coats

OR

Applying PU Based Coating – 2 Coats

- For Old Existing Chhajja where there is not any loose R.C.C./Plaster or having Tiling/Kota Stone on it, then the waterproofing system should be as follows:
- Applying Acrylic U.V. stable Coating with Fiber Net – 3 Coats

Step-4: Protection Layer

- Apply I.P.S./Screed/Brick Bat Coba/Plaster with non-woven geotextile as per site conditions and tile, and use cementitious grout.

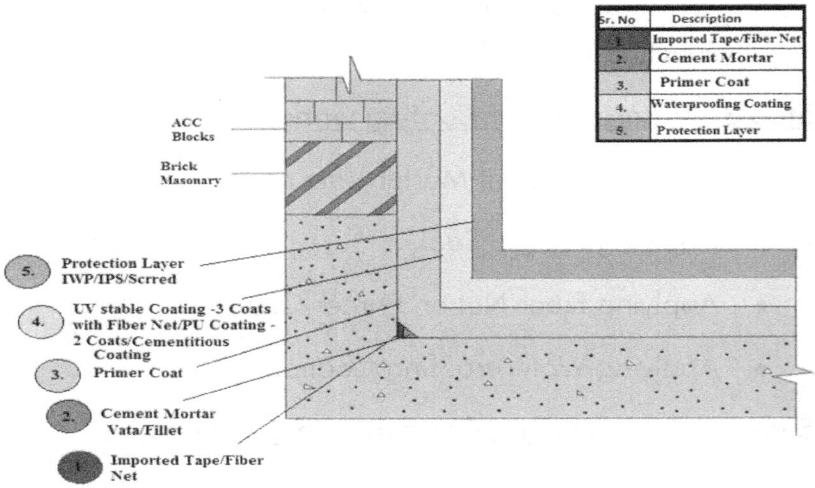

Fig. 6.4.2 Chhajja / Weather Shade Waterproofing

Please scan the Q.R. Code for further information:

"You cannot believe in God until you believe in yourself."

– Swami Vivekananda

6.4.3 Balcony Waterproofing

SYSTEM OF WATERPROOFING:

What is Balcony?

A balcony area is a extended area on up level floor which is useful for surrounding views & increase appearance of a building.

Importance of Balcony Waterproofing :

This area is indirectly & directly connected with weather i.e. monsoon & summer. So, this area must be well planned waterproofed to avoid internal damage, regular maintenance, mold & mildew, structural longevity, water ingress & for using external area at its fullest.

Step-1: Borehole Packing Work

Surfaces to be waterproofed shall be sound, clean, free of dust, paint, Curing agents, loose plaster debris etc.

All final cleaning will be carried out with an electric blower & Water jet (if required).

Finally, all the surface areas shall be dampened thoroughly with clean water, as this is necessary prior to the application of any type of coating.

Cracks shall be cut down by mechanically in "V" shape, cleaned & filled up with a paste of Polymer Modified Mortar or Crack- X Powder, Acrylic or P.U. Sealant, as per site condition [Refer Chapter 7.2].

Applying Polymer Modified Mortar (P.M.M.) @ Junction & Honeycombed Area With S.B.R.: Water: Cement.

Borehole Packing System: -

Applying Borehole Packing Work with Non-Shrink Free Flow Grout & Bath-seal Tape System + Geotextile with sandwich coating.

Step-2: Junction Area Treatment @ Floor & Wall

Cleaning Work as per requirement.

Applying crack filling/PMM work as per Site Conditions.

Applying Fiber net of 2 to 4 inch-wide sandwich coat or Imported Tape Joint System with Adhesive @ All Junction Area.

Step 3: Waterproofing Coating work

Applying Primer Coat + Fiber net + Two Coats of Two Component Flexible Cementitious Coating

OR

Applying Primer coat + Applying 3 Coats of Two Component Flexible Crack Bridging Waterproof Coating.

OR

Applying Primer coat + Applying PU Coating – 2 Coats

Applying 100 GSM Geotextile /Geotextile after Coating & Before IWP / IPS / Screed on Floor Area

Step-4: Protection Layer

Applying India Waterproofing (I.W.P.) / I.P.S. / Screeding with Integral Waterproofing Compound on Floor Area

Applying Plaster / P.M.M. with Integral Waterproofing Compound on Wall Area

Step-5: Applying for China Mosaic Work / Tilling Work

Tile with spacers 3 to 5 mm with tile adhesive as per site conditions & as per architect or interior designer with providing proper full slope @ down take pipe.

Sr. No.	Description
1	Imported Tape/ Fiber Net
2	Cement Mortar Vata/Fillet
3	Primer Coat
4	Cementitious Coating - 2 to 3 Coats with Fiber Net/ PU Coating - 2 Coats
5	Protection Layer IWP/IPS/Screed
6	Tiling Work with tile grout & tile on tile mortar

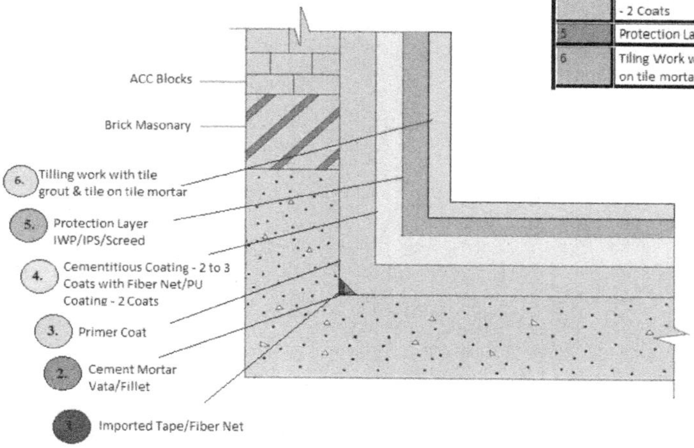

Fig. 6.4.3 Balcony Waterproofing

Please scan the Q.R. Code for further information:

"Science is a beautiful gift to humanity;
we should not distort it."

- A.P.J. Abdul Kalam

6.4.4 Open Terrace Space (OTS) Area Waterproofing

This is also called open to the sky. It is also called enclosed space within the building. This is because it is surrounded by buildings such as the living room, kitchen & bedroom from all sides.

Mainly, waterproofing is the process designed to prevent water from penetrating into the structure, so waterproofing is done in various layers and stages to create multiple barriers so that water cannot penetrate the structure. While waterproofing, a structure is waterproofed by the use of membranes and coatings. It is also called a "Building Envelope" waterproofing will give complete protection from weather, seepages from the ground and vertical travel of water in a structure.

SYSTEM OF WATERPROOFING

Step-1: Borehole Packing Work

Surfaces to be waterproofed shall be sound, clean, free of dust, paint, Curing agents, loose plaster debris etc.

All final cleaning will be carried out with a wire brush, pressurised water & an electric blower.

Finally, all the surface areas shall be dampened thoroughly with clean water, as this is necessary prior to the application of any kind of coating.

Cracks shall be cut down by mechanically in "V" shape, cleaned & filled up with a paste of Polymer Modified Mortar or Crack- X Powder.

Borehole Packing System

- Borehole Packing Work with Non-Shrink Free Flow Grout & Imported Tape Joint System/ Bituminous Tape/ Geotextile on (junction area- Pipe & Floor/ Wall Area) with required adhesive of sandwich coat + Plaster/P.M.M./Screed/Vata (if req.)
- Borehole filling in first stage (Up to half depth of slab approx.) filling by non-shrink free flow cementitious grout.
- Borehole filling is in the second and final stage (about half the depth of the slab to the floor).
- Finishing of the borehole from top and bottom, filled by Non-shrink free flow cementitious grout. (as required)
- After finishing apply with Imported Tape Joint System/ Bituminous Tape/Geotextile on (junction area-Pipe & Floor/Wall Area) with required adhesive of sandwich coat.
- P.M.M./Plaster/Screed/Vata @ wall & floor (if req.).

Step-2: Junction Area Treatment @ Floor & Wall

Cleaning Work as per requirement.

Applying crack filling/PMM work as per Site Conditions.

Applying vata @ floor & wall junction with styrene-butadiene rubber (S.B.R.) with fibre net with cement mortar.

Step-3: Waterproofing Coating work

Applying primer styrene-butadiene rubber (S.B.R.: Cement = 1:1.5) on perfectly dry prepared bed with a brush in a single coat @ 20-22 sq. ft per kg. The primer shall be allowed to dry for a minimum 2-3 hrs or used as per TDS (Technical Data Sheet) or PDS (Product Data Sheet).

Applying 40/60 GSM glass fiber net on primer coat.

Applying 1st Coat of Two Component Cementitious & Acrylic flexible coating shall be applied over the primed surface by brush.

Applying 2nd & final coat Two Component Cementitious & Acrylic flexible coating shall be applied. The coat shall be allowed to dry for 24-36 hrs.

Step-4: Protection Layer

For Floor Area:-

Laying I.P.S. / Screed Up to average. 1" to 1.5" with Integral waterproofing compound. (As per site conditions)

OR

Laying Brick Bat Coba with cement mortar mixed in, including mixing of Integral waterproofing compound on Floor Area.

For Wall Area:-

Apply finish coat plaster on the wall surface with an integral waterproofing compound on the wall area.

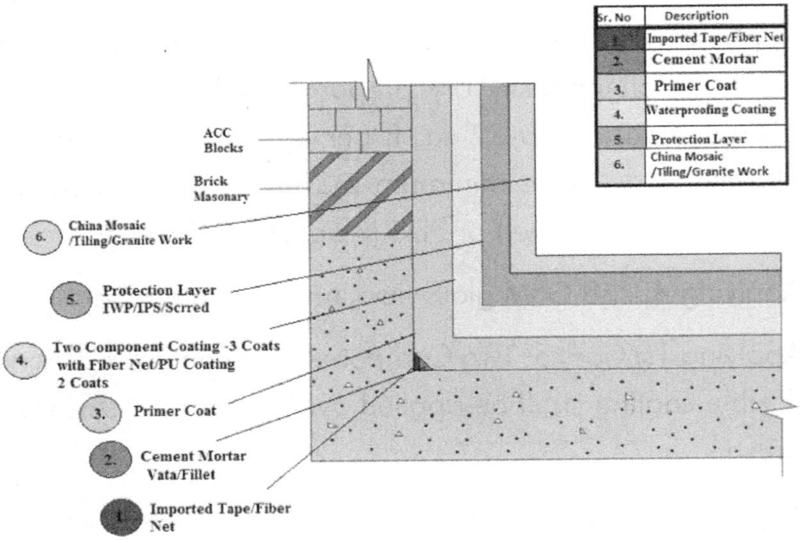

Fig. 6.4.4 Open Terrace Space/Open to Sky (OTS) Waterproofing Work

Thus, OTS/Open To Sky area is generally used or designed for ventilation, natural light, increasing space of living, creates a modern & sophisticated look, peaceful health for perfect & well designed Waterproofing System.

Please scan the Q.R. Code for further information:

"The purpose of life is a life of purpose."

- Robert Byrne

6.4.5 Industrial Terrace Waterproofing

Waterproofing of Industrial sheds is completely different from concrete waterproofing in the corrosion part occupying the metal surface is taken into consideration. It requires a different variety of waterproofing products like acrylic, cementitious, elastomeric, P.U.D., Epoxy and Polyurethane.

In India for different kinds of industries, different kinds of material used in Terrace, Cement Sheet Terrace, Metal (G.I.) Sheet Terrace, Prelaminated Sheet Terrace, R.C.C Terrace, Fiber Reinforced Polymer (FRP) Sheet, Polycarbonate Sheet Terrace, Green Roof Terraces & many more types of terraces.

Specification: System of Waterproofing

Step-1: Cleaning Work & Crack filling Work

Surfaces to be waterproofed shall be sound, clean, free of dust, paint, Curing agents, loose plaster debris etc.

All final cleaning will be carried out with an electric blower & Water jet (if required).

Finally, all the surface areas shall be dampened thoroughly with clean water, as this is necessary prior to the application of any type of coating.

Applying for Anchor Grout Filling Work as per site conditions

Step 2: Junction Area Treatment

Applying Junction of Floor & Wall Area treatment With

Sealing Tape with Adhesive @ Junction Area

Applying Cement Mortar Vata @ all Horizontal & Vertical Junctions with Integral Waterproofing Compound.

Step 3: Waterproofing Coating work

Applying U.V. Stable Acrylic Heavy Duty Waterproofing Coating work

(Cleaning + Self Penetrating Primer + Net/ Non-Woven Geotextile/ Imported Tape Joint (As per requirement) + U.V. Stable Acrylic Heavy Duty Waterproof / Bituminous Coating / P.U. Coating - 3 Coats) = 5 Layers System @ Junction Area)

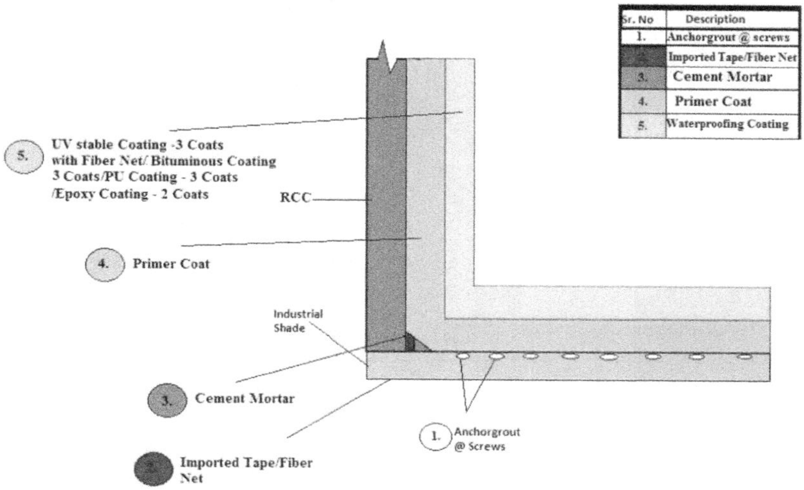

Fig. 6.4.5 Industrial Terrace Waterproofing

Also As per chemical used/developed in the industry. Waterproofing designed is to be prepared accordingly with a team work of client, structural consultant, Architect, Waterproofing manufacture company & Waterproofing Consultant, to better service life of terrace.

Please scan the Q.R. Code for further information:

"Arise! Awake! And stop not until the goal is reached."

- Swami Vivekananda

6.5 General Area Active Leakage Stopping by Injection Grouting

This is the method of filling the cracks, open joints, voids or honeycombs in concrete or masonry structural members. This is done with a grout material that gives desired results like strengthening a structure and preventing water ingress.

6.5.1 Cementitious Injection Grouting

It is used where strength enhancement of structure is not required. This is popular because it is readily available in the market. This method requires high water and cement content for injection purposes.

Cement Grouting is preferred because it is a homogeneous material, and a low budget is required.

Application:

- Remove the existing damaged/broken/loose R.C.C. area.
- Drilling 12 mm diameter holes with electric drilling machines along the honeycombed area, and if required @ floor wall joints slab and fixing Nozzles in place using quick setting compound
- Material (Cement, Expanding grout additive, S.B.R.) Mix as required.
- Providing & Filling Polymer Modified Mortar in Honey

Combed R.C.C./Brick Masonry Area after finalising & fixing the Nozzles. [if required]

- Pressure grouting should be carried out by using Cementitious Hand operated Cementitious Grouting Pump through the Nozzles, starting from one Nozzle and stopping the operations when the grout starts coming out from an adjacent Nozzle; this is done similarly in all the Nozzles.
- After Completion of Cementitious Grouting work, Remove or Cut the Nozzles & Plug the holes filled by Non-Shrink Free Flow Cementitious Grout/ Repair Mortar/ Polymer Modified Mortar (P.M.M.).
- Make curing or apply curing compound to that area.

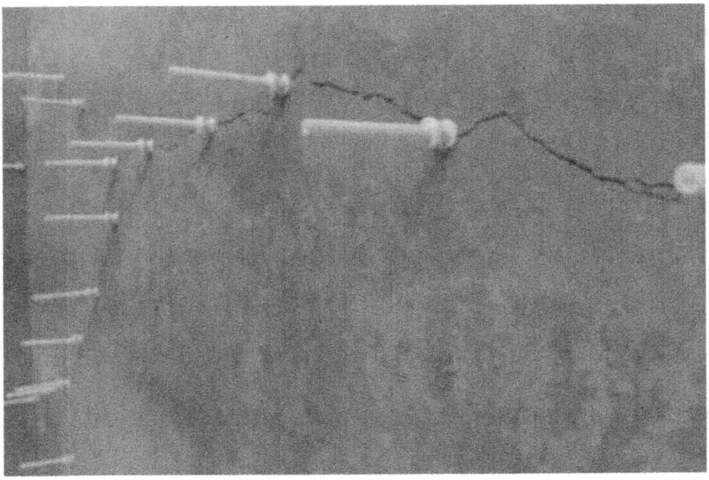

Fig. 6.5.1 Cementitious Injection Grouting Work

Please scan the Q.R. Code for further information:

"Genius is one per cent inspiration and ninety-nine per cent perspiration."

\- Thomas Edison

6.5.2 Acrylic Injection Grouting

This is used for:

1. Underground structure (such as Subway, Tunnel, Parking lot, Manhole etc.)
2. Civil structures such as Dam, etc.
3. Concrete structure.
4. Leaking area of underground structure.
5. Constructed joint area and cracked area.

Application:

- Drill holes: Check the cracks and drill holes to the rear side of the concrete structure, and clean the holes by air-compressor.
- Installing packers: Insert packers into holes and fasten packers tightly using a T-wrench.
- Pour an agent into Tank A
- Add catalyst at an agent, and mix each other using a stick.
- Pour B agent into Tank B.
- Checking hardening time

(a): pour each agent into each paper cup.

- Checking hardening time.

(b): mix together with a 1:1 Mix ratio

- Checking hardening time.

(c): If the hardening time is long, increase the catalyst so as to shorten it.

- Connect hoses to pistons.
- Connect the acrylic gun to the hose.
- Connect the acrylic gun with installed packers.
- Turn the power switch on to run the machine, and inject the acrylic resin into every hole.
- After finishing the injection, remove the packers using a hammer.
- Clean the holes/crack areas and coat the sealing material.

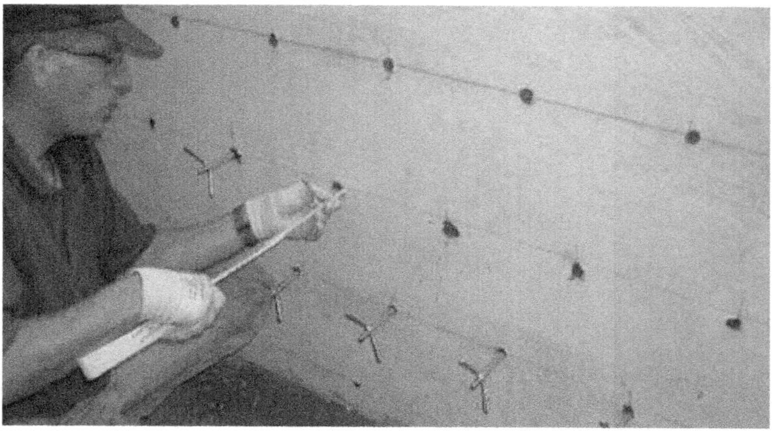

Fig. 6.5.2 Acrylic Injection Grouting Work

Please scan the Q.R. Code for further information:

"Life is like riding a bicycle. To keep your balance, you must keep moving."

- Albert Einstein

6.5.3 Poly Urea/ Poly Urethane Injection Grouting (PUIG)

It is used where concrete slabs and walls are used as the environmental barrier. Here, Water leakage is a problem.

On diagnosed a leakage issue following five classic approaches to an acceptable resolution can be explored with the assistance of water proofing expert/consultant.

After a leakage problem has been diagnosed, five classic approaches to an acceptable resolution can be explored.

- Crack/Joint Routing, Caulking, and/or Dry-packing
- Crack/Joint Injection, Chemical Grouting
- Water Management & Drainage
- Coatings, Sealers, Reactants, Sheet Liners
- Electro-Osmotic Pulse (E.O.P.) Technology

Through PU technique or P.U. injection grouting is a waterproofing technique to inject polyurethane to prevent water leakage through cracks or gaps in the building or concrete. This is immediate stop activity with a low viscosity and it is a key to seal the small cracks.

Application:

- Demolition or Removal of existing damaged/broken/ loose R.C.C. area.

- Drilling 14mm diameter holes with electric drilling machines along the honey combed area and if required @ floor wall joints slab and fixing Aluminium or G.I. N.R.V. heavy nozzles of 10 mm diameter in place using quick setting compound.
- Filling Polymer Modified Mortar in Honey Combed R.C.C./Brick Masonry Area after finalising & fixing the Nozzles.
- Pressure grouting should be carried out by using [PU FOAM] Hand operated Centrifugal PU grouting machine through the nozzles, starting from one nozzle and stopping the operations when the grout starts coming out from the adjacent nozzle; this is done similarly in all the nozzles. [As per client's suggestion or actual site condition requirement]
- After Completion of P.U. Grouting work, Remove or Cut the Aluminium or G.I. N.R.V. heavy nozzles & plug the holes filled by quick setting compound or Non-Shrink Free Flow Cementitious Grout/ Repair Mortar /Polymer Modified Mortar (P.M.M.).
- Providing & Applying Crystalline Waterproofing -2 Coats on only the Nozzles area.
- Make curing or apply curing compound to that area. [If required]

Fig. 6.5.3 P.U. Injection Grouting Work

Please scan the Q.R. Code for further information:

"Education is not preparation for life.
Education is life itself."

- John Dewey

6.6 Dam Waterproofing

Dams are playing essential role to store water for different purposes like irrigation, water supply hydroelectric power and flood control.

The Structure of dams are full of Complexities, only specialists can understand and give problem solution or advice to safe guard it.

Typical Problems in dams are as follows,

- Cracks in the dam body
- Leakage in the intake gate
- Leakage in draft tube
- Leakage in the inspection gallery
- Abrasion of spillways
- Cavitations in peripheral walls
- Damages to baffle blocks
- Abrasion and impact damages on the spillway and the stilling basin

Concept of the waterproofing layer (Protection layer):

After studying the exact condition, the expert will decide which waterproofing system is to be applied. i.e. (1) Support system (Which includes Drainage layer, Protection layer, filter layer, etc.)

(2) Protection system (Synthetic Protection, Mineral Protection, combination, etc.) instruction of Geotextile Geo membrane or Geo composite will be decided by the waterproofing expert consultant.

During Dam Waterproofing, we must care, think & implement about drainage system, monitoring system, standard compliance, regular construction practices ,environmental conditions for future, selection of material & perfect application & Annual maintenance regularly.

In India for new dams & existing dams crores of rupees invested every year in waterproofing.

So, Dam Waterproofing is must for structural safety, longevity, erosion control & leakage prevention.

Please scan the Q.R. Code for further information:

"Dams have harmed our wildlife and made rivers less useful for reaction."

- Stephen Ambrose

6.7 Tunnel Waterproofing

It is again a challenging job which cannot be performed without an expert's advice. Damage from hydrostatic pressure, corrosive chemicals carried by groundwater and extreme temperature fluctuations will hamper the tunnel's structural integrity. For example, many tunnels are pouted Under Sensitive locations where any failure can lead to substantial repair expenses & disruption of essential activities. Further tunnel repair & waterproofing may not even be possible without removing tons of overburden.

Wood class Water Proofing:

Tunnel waterproofing technologies prevent water penetration to the tunnel interior. Successful waterproofing of tunnel includes:

- Smoothing shotcrete
- Anchoring system
- Waterproofing membrane
- Induction Welding Plates
- Water Barriers
- Tunnel grouting system, etc

Selection criteria of a satisfactory tunnel waterproofing system:

Here, it is necessary to have an in-depth knowledge of the

principles of waterproofing below-ground structures. The following points are necessary to be kept in mind when selecting a satisfactory Waterproofing System.

- Type of structure
- Type of construction (bored, cut-and-cover, etc.)
- Depth of burial of structure (lowest elevation of structure)
- Maximum Elevation of groundwater (after construction)
- Environmental conditions at the site (including pH, hydrocarbons, and chemicals)
- Ease of installation
- Track record (successes and failures of the system)
- Construction Cost
- Long term maintenance

Different types of tunnel waterproofing systems:

Out of the various waterproofing systems available, some of the important ones are mentioned below.

- Liquid Waterproofing System
- Panel Waterproofing System
- Membrane Waterproofing System
- Sheet Membranes Waterproofing System

- The epoxy liner-proofing system
- The spray-on waterproofing membrane system
- Cementitious waterproofing system
- Hybrid Waterproofing System

In India every year crores of rupees are invested in waterproofing of new tunnels & maintenance of existing tunnels, so tunnel waterproofing is very crucial & important.

The waterproofing consultant will decide what type of system or combination of systems to be used for tunnel waterproofing.

Please scan the Q.R. Code for further information:

"The light at the end of the tunnel is just the light of an oncoming train."

- Robert lowell

6.8 Canal Waterproofing

Why Canal Waterproofing is necessary?

Canal Waterproofing is required in construction due to mentioned reasons,

- For Potection of environment
- For Support of Agricultural irrigation project
- For Standard Compliance
- For decreasing maintenance cost
- For increasing life cycle

INSTALLATION OF CANAL LINING

It is necessary to study the actual conditions under which the waterproofing system has to be installed & has to work. Different parameters of waterproofing the canal will lead to a malfunction in the system. Do geological & geotechnical conditions have to be investigated or consist of investigated on-site

WATERPROOFING SYSTEM:-

Thus, the waterproofing system must consist of,

1. Support
2. Drainage Layer, which includes Protection Layer & filter Layer.
3. Waterproofing Layer.

4. Protection, which includes synthetic Protection, Mineral protection & combination of both.

In India for new Canals & existing Canals crores of rupees invested every year in waterproofing.

So, canal Waterproofing is also highly technical and only with the assistance water proofing consultant this can be made successful.

Please scan the Q.R. Code for further information:

"Many of life's failures are people who did not realise how close they were to success when they gave up."

- Thomas A. Edison.

Chapter

7

Structural Strengthening - Retrofitting/ Repairing Work

7.1 Structural Repairing FAQs

Q-1: What is Structural Repairing Work?

A-1: In Construction industries, Structural Repairing Work are a special technical expertise work for old existing damage & new poor construction work to repair & strengthen it for fulfilling the purpose of service life of buildings/structures.

Q-2: Why is structural repairing work / structural strengthening required today?

A-2: Structural Repairing Work / structural strengthening is required because of the following reasons:

1. Due to poor maintenance in old existing damaged buildings/structures
2. Due to poor workmanship in new-running construction work
3. Due to changes in the Government policy for buildings/ structures bylaws
4. Due to poor waterproofing work, the structure may get damaged internally.

Q-3: What are the methods or systems used for structural strengthening or repairing works?

A -3: There are various methods or systems used for structural strengthening or repairing works as below,

1. Systematic Repair
2. Epoxy Injection Grouting
3. Ready to use mortar
4. Fiber Reinforced Polymer

a) Glass Fiber Reinforced Polymer

b) Carbon Fiber Reinforced Polymer

c) Aramid Fiber Reinforced Polymer

Please scan the Q.R. Code for further information:

"Life is a succession of lessons which must be lived to be understood."

- Ralph Waldo Emerson.

7.2 Crack Repair

A crack is a complete or incomplete partition of concrete into two or more parts.

Mainly Two (2) types of cracks in the structure:

(1) Structural Cracks

(2) Non-structural Cracks

Cracks are also depends on actual materials, based on appearance, movement & location.

Cracks Repairing Treatment

Step-1 Cleaning & crack filling

- Firstly, basic cleaning work is required as per site conditions.
- Cracks shall be mechanically cutting in "V" shape, cleaned it & filled up with a paste of Polymer Modified Mortar (P.M.M.) or Crack- X Powder as per Site Conditions/requirement.
- Applying Fiber Net /Mesh or Non-Woven Geotextile on it.

Step-2 Epoxy Injection Grouting

- Removal of existing damaged/broken/loose R.C.C. area
- Drill 12 mm diameter holes with electric drilling

machines along the honeycombed area, if required, about the floor wall joint slab, and put in place nozzles using a quick-setting compound.

- After finalising & fixing the Nozzles, providing & filling Polymer Modified Mortar (P.M.M.) in Honeycomb R.C.C./Brick Masonry Area. [if required]
- Pressure grouting should be carried out using an epoxy grouting pump starting from one nozzle and stopping the operations when the grout starts coming out from the next nozzle. This is done in the same way for all the nozzles.
- After Completion of Epoxy Grouting work, eliminate or cut the nozzles & plug the holes filled by Insta-stop or Non-Shrink Free Flow Cementitious Grout/ Repair Mortar/ Polymer Modified Mortar (P.M.M.).
- Applying curing compound on that area.

Please scan the Q.R. Code for further information:

"The only impossible journey is the one you never begin."

- Tony Robbins

7.3 Anti Corrosive Treatment

Corrosion:

Corrosion is a natural phenomenon that transforms a pure metal into a more chemically steady form such as oxide, hydroxide, or Sulphide. It is the slow destruction of materials by chemical or electrochemical reactions with their environment.

- Corrosion mainly impacts structural safety, Health & environmental issues.
- India has 5-6% of losses annually on account of corrosion. (More than Seven Lakh Crores)

Causes of corrosion:

1. Presence of impurities in the atmosphere
2. Temperature
3. Humidity
4. pH

Structural effects due to corrosion:

1. Reduced Steel – Concrete bond
2. Limited Ductility (Reduced Ductility)
3. Loss of Strength
4. Fatigue

Types of Corrosion:

1. Microbial corrosion
2. Crevice Corrosion
3. High Temperature Corrosion
4. Galvanic Corrosion
5. Pitting Corrosion

Corrosion Rate:

In Rural environment areas, the Corrosion Rate is 4-60 μm.

In an Urban environment area, the Corrosion Rate is 30-70 μm.

In the Industrial environment area, the Corrosion Rate is 40-160 μm.

In the Marine environment area, the Corrosion Rate is 60-170 μm.

Treatment/ Methodology:

Step - 1: Surface Preparation as per ISO 8501-1,

(i) S.A. 1 – Light Blast Cleaning

The jet passes quickly over the entire surface. Therefore, loose mill scale, rust & foreign matters are removed.

(ii) S.A. 2 – Through Blast Cleaning

SA-1 is done similarly, after which the surface is cleaned with a vacuum cleaner, clean & dry compressed air or cleaned with a brush.

(iii) S.A. 3 – Very through Blast Cleaning

The mill Scale, rust & foreign matter shall be removed to such an extent that only remnants appear as shades on the surface.

(iv) S.A. 4 – Blast Cleaning to visually clean steel

Cleaning work to be done as per up to S.A. 3. Thereafter, it should be clean with dust free. Therefore, the surface should have properly clean for any type of coating.

<u>**Step - 2:**</u> After Proper Cleaning Work, apply epoxy-based primer – E.P. primer.

<u>**Step – 3:**</u> After drying E.P. primer, applying Two Component Zinc rich primer coating (As per BS 4652, Type – 2 for chloride induced).

Please scan the Q.R. Code for further information:

"The purpose of our lives is to be happy." - Dalai Lama

7.4 Systematic Repair

Main Causes:

1. Structural Damage
2. Chemical Damage
3. Corrosion of Reinforcement
4. Fire damage
5. Seismic damage
6. Overload
7. Insufficient Reinforcement

PROCEDURE

System of Strengthening by Systematic Repair

Applying breaking & removal of loose plaster work as per requirement.

Applying cleaning work as per requirement.

Applying Epoxy Injection Grouting from Ceiling/Terrace Top Side if through & through cracks are found.

Applying Rebaring where steel is broken. (If required)

Applying rust remover/rust convertor on the steel reinforcement bars is quite important for the bonding between reinforcement & high-strength mortar.

Applying zinc-rich primer (Anti-carbonation coating) to the

steel bar to prevent corrosion of steel reinforcement [After approx. Half an hour].

Applying a primer coat of any SBR/LATEX/Acrylic @ removal area to provide sufficient bonding between old existing concrete & high strength mortar.

Applying plaster/ levelling with cement sand aggregate when depth is > 50 mm @ removal area in stage 1 as per site condition.

Applying plaster/ levelling with cement sand aggregate when depth is > 30 mm @ removal area in stage 2 as per site condition. [2nd day]

Applying plaster/levelling (cement & sand mortar) at removal area to give final finishing as per site condition.

Applying curing compound @ removal area after the finishing in about half an hour.

Sr no	Description
1.	Epoxy Injection Grouting/ Rebaring
2.	Rust Converter/Rust Remover
3.	Zinc rich Primer
4.	SBR/Latex/Acrylic
5.	Cement:Sand:Aggregate
6.	Plaster (Cement:Sand)

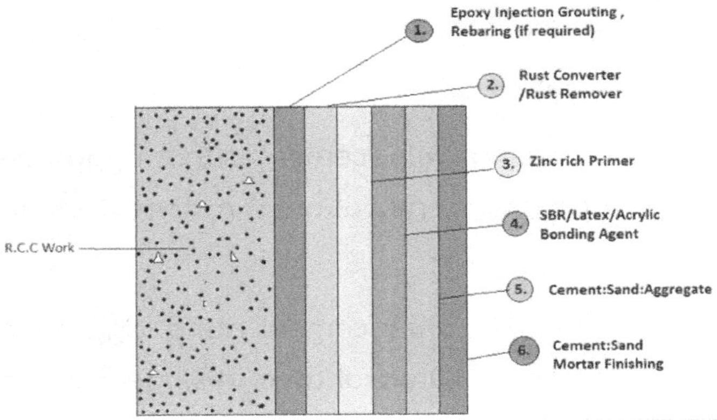

Fig. 7.4 Systematic Repair

Please scan the Q.R. Code for further information:

"The greatest glory in living lies not in never falling, but in rising every time we fall."

- Nelson Mandela

7.5 Strengthening

Strengthening of structure is the process of upgrading a structure to improve performance under existing loads or to increase the strength of structural members to carry additional loads.

There are many methods available. The most used methods are as follows:

7.5.1 Epoxy Injection Grouting

- Removal of existing damaged/broken/loose R.C.C. area
- Drilling 12 mm diameter holes with electric drilling machines along the honeycombed area, if required, @ floor wall joints slab, and fixing Nozzles in place using the quick setting compound.
- Providing & filling Polymer Modified Mortar in honeycombed R.C.C./Brick Masonry Area after finalising & fixing the nozzles. [if required]
- Pressure grouting should be carried out using an Epoxy Grouting Pump through a Nozzle starting from one nozzle and the operation should be stopped when the grout starts coming out of the adjacent nozzle, this is done in the same way in all nozzles.
- After Completion of Epoxy Grouting work, Remove or Cut the Nozzles & Plug the holes filled by Insta-stop

or Non-Shrink Free Flow Cementitious Grout/ Repair Mortar/ Polymer Modified Mortar (P.M.M.).

- Make curing or apply curing compound to that area.

Please scan the Q.R. Code for further information:

"Life is really simple, but we insist on making it complicated."

\- Confucius

7.5.2 Ready to use mortar

System of Strengthening by Ready-to-Use Mortar

Applying breaking & removal of loose plaster work as per requirement.

Applying cleaning work as per requirement.

Applying rust remover/rust convertor on the steel reinforcement bars which is quite important for the bonding between reinforcement & high strength mortar.

An anti-carbonation coating on the steel bar is applied to prevent corrosion of steel reinforcement.

Applying primer coat of any SBR/LATEX/Acrylic @ removal area to provide sufficient bonding between old existing Concrete & high strength mortar.

Applying plaster/ levelling (high strength ready use mortar) @ removal area in stage 1 as per site condition & as per TDS (Technical Data Sheet)/ PDS (Product Data Sheet). [2nd Day]

Applying plaster/ levelling (high strength ready to, use mortar) @ removal area in stage 2 as per site condition & as per TDS (Technical Data Sheet)/ PDS (Product Data Sheet). [3rd Day]

Applying plaster/levelling (high strength ready use mortar) @ removal area in stage 3 as per site condition & as per TDS (Technical Data Sheet)/ PDS (Product Data Sheet). [4th Day]

Applying plaster/levelling (cement & sand mortar) removal area to give final finishing as per site condition & as per TDS (Technical Data Sheet)/ PDS (Product Data Sheet). [5th Day]

Applying curing compound @ removal area after the finishing in about half an hour.

Fig. 7.5.2 Ready to use Mortar

Please scan the Q.R. Code for further information:

"Life is either a daring adventure or nothing."

- Helen Keller.

7.5.3 Fiber Reinforced Polymer (F.R.P.)

Fiber-reinforced polymer is made of a protective polymer reinforced with high-strength & lightweight fiber composition.

TYPES OF FRP =>

1. Glass Fiber Reinforced Polymer (GFRP)
2. Carbon Fiber Reinforced Polymer (CFRP)
3. Aramid Fiber Reinforced Polymer (AFRP) [Short form of Aromatic Poly Amid]

7.5.3.1 Glass Fiber Reinforced Polymer

System Methodology

- Applying breaking/grinding/chamfering up to the mother surface
- Apply minute cleaning with a wire brush to remove dust.
- Applying Levelling with Epoxy putty & filling every void with epoxy filler.
- Applying Epoxy Base Primer.
- Apply Epoxy Base Adhesive on a primer coat.
- Stick Glass Fiber perfectly without gap & apply Roller on the Glass Fiber [unidirectional or bidirectional as per requirement of the structure]

- Applying 2nd Coat of Epoxy Base Adhesive before plaster.
- Applying Plaster for Finishing.

Sr no	Description
1	Epoxy puty
2	Primer (Epoxy Based)
3	Adhesive (Epoxy Based)
4	Glass Fiber
5	Adhesive (Epoxy Based)
6	Plaster

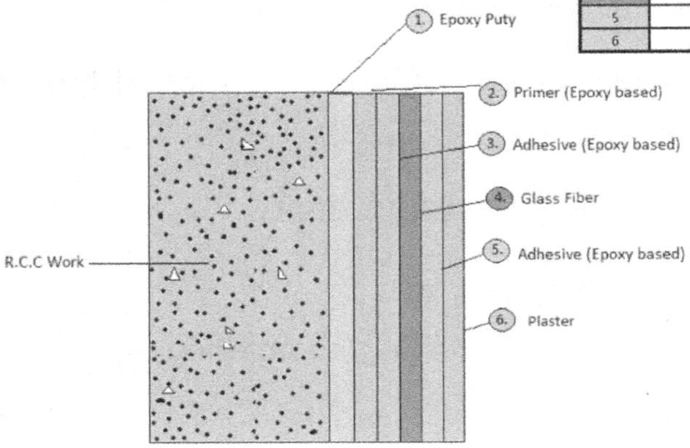

Fig. 7.5.3.1 Glass Fiber Reinforced Polymer

Please scan the Q.R. Code for further information:

"Always remember, your focus determines your reality."

- George Lucas

7.5.3.2 Carbon Fiber Reinforced Polymer

System Methodology

- Applying breaking/grinding/chamfering up to the mother surface
- Applying minute cleaning with a wire brush to remove dust.
- Applying Levelling with Epoxy putty & filling every void with epoxy filler.
- Applying Epoxy Base Primer.
- Apply Epoxy Base Adhesive to a primer coat.
- Stick Carbon Fiber perfectly without gap & Apply Roller on the Carbon Fiber [unidirectional or bidirectional as per the requirement of the structure]

- Applying 2nd Coat of Epoxy Base Adhesive before plaster.
- Applying Plaster for Finishing.

Sr no	Description
1	Epoxy puty
2	Primer (Epoxy Based)
3	Adhesive (Epoxy Based)
4	Carbon Fiber
5	Adhesive (Epoxy Based)
6	Plaster

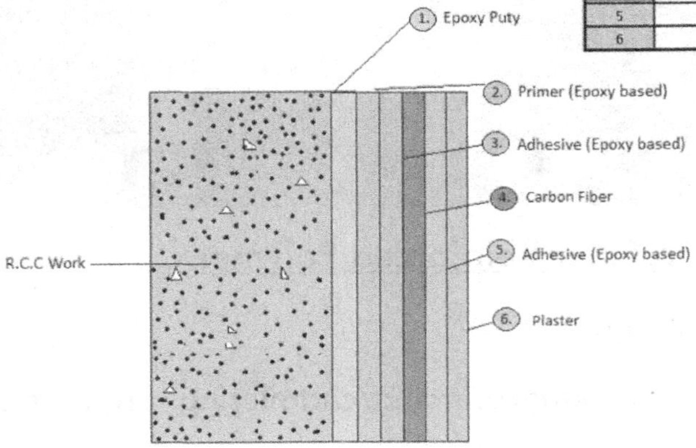

Fig. 7.5.3.2 Carbon Fiber Reinforced Polymer

Please scan the Q.R. Code for further information:

"Leave nothing for tomorrow which can be done today."

- Abraham Lincoln

7.5.3.3 Aramid Fiber Reinforced Polymer (AFRP)

Why AFRP used in construction Industry:-

AFRP is mainly used in our industry due to following reasons:-

- Increase in Sustainability
- Due to its high strength to Weight Ratio
- Flexibility
- Retrofitting during seismic conditions
- Performance increase
- Durability increase
- During application minimal disruption
- Very Good electrical & thermal properties

System Methodology

Applying breaking/grinding/chamfering up to the mother surface

Applying minute cleaning with a wire brush to remove dust.

Applying levelling with Epoxy putty & filling every void with epoxy filler.

Applying Epoxy Base Primer.

Apply Epoxy Base Adhesive to a primer coat.

Stick Aramid Fiber perfectly without gap & Apply Roller on the Aramid Fiber [unidirectional or bidirectional as per the requirement of the structure]

Applying 2nd Coat of Epoxy Base Adhesive before plaster.

Applying Plaster for Finishing.

Sr no	Description
1	Epoxy puty
2	Primer (Epoxy Based)
3	Adhesive (Epoxy Based)
4	Aramid Fiber
5	Adhesive (Epoxy Based)
6	Plaster

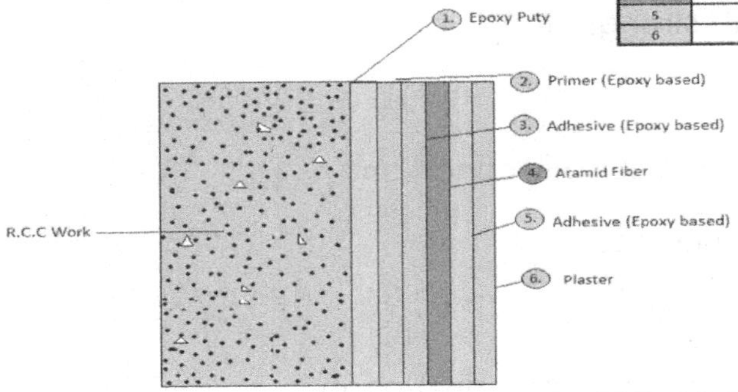

Fig. 7.5.3.3 Aramid Fiber Reinforced Polymer

Please scan the Q.R. Code for further information:

"Before anything else, preparation is the key to success."

- Alexander Graham Bell.

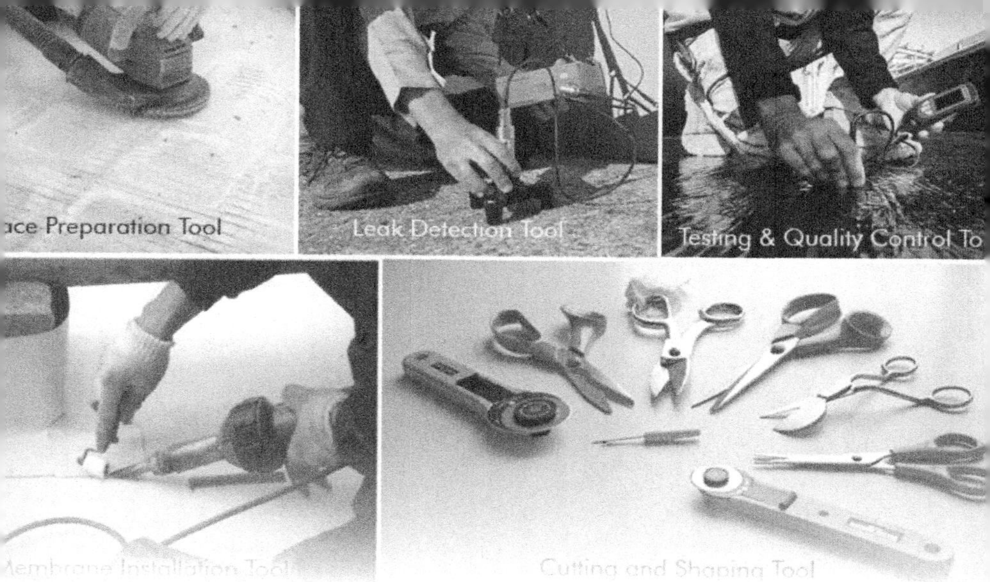

Chapter

8

Modern Tools Used in Waterproofing Industry

Up-till now, for doing waterproofing only traditional tools like Sprayers, Pressure Washers, Brushes & Rollers, Heat Guns and Torches, Sealant Guns, Trowels, Measuring Tools, Drills and Mixers etc. were used. The benefit of there tools were temporary & you have to spend money after sometime for repairing etc. This was resulting in lot of physical & financial difficulties.

Now, lot of developments have taken place during last 5 to 10 years & waterproofing system have observed sea change. National & International waterproofing standards & regulations have also improved. To get benefit of

1) Cost effectiveness
2) Data Driven Decision Making
3) Improved Efficiency & Accuracy
4) Regulatory Compliance
5) Adaptability to complex projects
6) Innovation and Competitiveness
7) Environmental Benefits
8) Enhanced Durability

9) Enhanced Safety & to finish the work speedily accurately & as per the standard of National & International authorities, the utilization of modern tools in waterproofing work is a must.

The following types of advanced tools are used in waterproofing industry.

1. Waterproofing Additives and Admixtures

    i) Crystalline Waterproofing Admixtures

    ii) Integral Waterproofing Compounds

2. Green Roof and Garden Waterproofing

    i) Root Barriers Membranes

    ii) Drainage Layer Systems

3. Spray - Applied Waterproofing Membranes

    i) Polyurea Spray Systems

    ii) Liquid Applied Membranes

4. Sheet Membrane Application Tools

    i) Heat Welders

    ii) Pressure Rollers

5. Advanced Testing and Quality Control

    i) Moisture Meters and Hygrometers

    ii) Adhesion Testers

6. Injection Grouting Systems

    i) PU (Polyurethane) Injection Pumps

    ii) Epoxy Injection Systems

7. Surface Preparation Tools

    i) Concrete Scarifiers and Grinders

    ii) Hydro Blasting Equipment

8. Specialized Coatings and Sealants

    i) Elastomeric Coatings

    ii) Silicone and Polyurethane Sealants

9. Leak Detection and Monitoring Tools

    i) Electronic Leak Detection (ELD) Systems

    ii) Infrared Thermography

10. Waterproofing Sheets and Panels

    i) Bentonite Waterproofing

    ii) Pre-formed Membrane Panels

Thus, These tools & technologies help ensure that waterproofing applications in various structures including residential, commercial & industrial buildings are effective, durable & reliable & one must use these tools for getting the best waterproofing results.

Please scan the Q.R. Code for further information:

"If you get tired, learn to rest, not to quit."

- Banksy

Chapter

9

Main Case Studies in Waterproofing

Case Study-1 (Rising Dampness/Capillary Action): There is a big builder in Ahmedabad. During the construction of his privately owned bungalow, he had already done the waterproofing from another waterproofing agency. The waterproofing has gone well for a while. However, after about 1 years, some walls showed rising dampness and leakage problems in some places. Due to this, the owner and other family people started to get disturbed. He called the agency, He had the work done and got it checked and then repaired. However, it did not last long; after after only 4 months, the problem of leakage started. After that, he called many other agencies and got it checked and repaired, but after 6 to 8 months, the problem of leakage started in that area & nearby areas. Due to this, their expenses increased. After That, they contact us. After this, we did a site visit and checked and found the main leakage point. After that, we came up with our 21-day method based Programme along with our decades of experience, which is the best way to solve it, and we got the job done accordingly. After this, we checked them using the conventional pond test by sprinkling continuously for 6 to 8 hrs at 7 days to determine whether there is a problem or not. Until now, after even more than

12 years, there has been no leakage problem. That is, if you already work with a good waterproofing method, it is a long-term investment for you that saves your other expenses in the future.

Please scan the Q.R. Code for further information:

Case Study-2 (Ceiling & Wall Area): The builder prepared tenders for waterproofing work from many agencies during the construction time of the new building. We also filled the tender and sent it to the builder. Our tender was higher according to rate and more systemize as compared to other agencies, Due to which the builder declined us and finalised to another agency for that work. The agency has started the work as per the finalisation of the work order. Thereafter, the agency completed the work as per the given timeline. After completion of approx. 1 years, in some bungalows, leakages were found in the ceiling & another wall area. So, the client requested that builder to resolve the problem as soon as possible. Thereafter, the Builder informed the waterproofing agency to resolve the problems. The agency tried many times to resolve the leakage issues, but due to some unwanted reasons, they have not achieved the required results for that issue. So, the builder contacted many agencies to resolve it.

Nevertheless, they have tried many other agencies, but the results were not obtained. Finally, the builder contacted us to resolve the leakage issues. Thereafter, we visited the site, found the root cause of the leakages, and finalised our best system for repairing that area, which is the best way to solve it. We got the job done accordingly. After that, we checked them by pond test to see whether there was a problem or not. after even 7 years, there is no leakage problem in that area. Therefore, if you already work with a consultant & good & futuristic waterproofing method, it is not at all an expenses, it's your investment for leak-proof & healthy peaceful life.

Please scan the Q.R. Code for further information:

Case Study-3 (Terrace, Swimming pool & Water Tank Area): One of the best architects were planning to prepare a new private bungalow (Own Bungalow) from foundation to finishing. Therefore, he met a structural consultant, project management consultant & builder for this work. Moreover, he discussed with each other & finalised the overall plan from the foundation to finish. Thereafter, the overall work starts from the foundation, as discussed with the structural consultant, project management consultant & builder. The builder completed the work in mutual discussion with the consultant and the architect, and he got the waterproofing done by his local agency using a conventional method. After completion of only 2.5 years in a bungalow, leakages were found in the terrace, water tank area & swimming pool area. So, the architect requested that builder to resolve the problems as early as possible. The builder contacted many other agencies to resolve the problems. The agencies tried many times to resolve the leakage issues, but due to some technical reasons & less experiences, they did not achieve the required results for that area. Thereafter, the builder contacted us by a referral from his friend. After that, we held meetings at the bungalow & discussed the solutions for leakages in the building. After that, we finalised our system for repairing that area, which is the best practical way to do it, and we got the job done accordingly. After even more than 10 years, there is no leakage problem in that area. Therefore, waterproofing work must be done using the latest methods, not by conventional methods. If you have already worked with a good expert, knowledgeable & experiences waterproofing consultant, you can live leak-proof, stress free, healthy, last long life as far as waterproofing concern.

Please scan the Q.R. Code for further information:

Case Study-4 (Basement & Retaining Wall Area): The Builder constructed a new building/scheme having -2 basements. According to their thinking, no waterproofing is required in the basement and retaining wall area. Therefore, they did not go with any waterproofing treatment in the basement area and retaining wall area. Buildings and other areas were treated with waterproofing treatment except the basement & retaining wall area. After completion of only 1.5 years, leakages were found in the basement & retaining wall area in the rainy season. The basement area was full of water up to 4" to 5" from the floor area. Therefore, the Lift area was blocked due to water travelling in that area. So, the client was complaining the builder to resolve this matter. Thereafter, the builder checked it and informed the project in charge to resolve this matter. The project in charge discussed with us. We met on site and found out the root cause of the water leakage, and finally, we reached out to water leakages due to the retaining wall not being properly treated and the nearby sewerage pipeline. Then, we designed our best system with the best waterproofing chemical coating & also injection grouting work in that area. That was the best way to solve it, and we got the job done accordingly. After even 15 years, there is no leakage problem in that area. Waterproofing work must be done in all areas as per requirement. If you have already worked with a good waterproofing consultant from start to finish, it is a long-term investment, not an expense.

Please scan the Q.R. Code for further information:

<u>Case Study-5 (Strengthening/Retrofitting Work):</u> The Builder was constructing a new commercial building. After some time, i.e. after approx. 5 years, the structural engineer raised the requirement to increase 4-5 floors in the building. To increase the floor of the building, they have to increase the strength of the column. Therefore, builders were contacted and informed of what to do to increase the structural strengthening of the column. Thereafter, we have suggested the best system for increasing the structural strengthening of the columns. Thereafter, we finalised the agency/contractor, gave instructions as per the design system, prepared a bar chart for the same and worked accordingly. We got the job done accordingly. After that we have saved other expenses for the reconstruction of the column area or building.

Please scan the Q.R. Code for further information:

For understanding the benefits of Waterproofing Case Studies are very important, Case Studies are also useful for minimizing the risks, ensuring durability & demonstrate the effectiveness of waterproofing solutions.

Please scan the Q.R. Code for further information:

"Experience is the teacher of all things."

- Julius Caesar

Chapter 10

Role of Coach in Life

The basic role of a coach in life is to support, guide, and empower individuals to reach their full potential and achieve their goals. Whether in sports, personal development, career, or any other aspect of life, a coach serves as a mentor, motivator, and strategist. They help individuals identify their strengths and weaknesses, set realistic goals, develop action plans, overcome obstacles, and stay accountable to their commitments. A coach also provides encouragement, feedback, and perspective, helping individuals navigate challenges and grow personally and professionally. Ultimately, the aim of a coach is to facilitate positive change and facilitate the journey of self-discovery and improvement.

Let's delve into the various aspects of the role of a coach in life:

1. **Strengths Identification**: Coaches help individuals recognise and leverage their unique strengths and talents. By conducting assessments and reflective exercises, coaches facilitate self-awareness and help clients identify their core competencies. Understanding one's strengths allows individuals to approach challenges with confidence and

capitalise on opportunities for growth and success.

2. **Skill Building and Development**: Coaches facilitate skill development and learning opportunities that empower individuals to enhance their capabilities and achieve their goals. Whether it's communication skills, time management, leadership abilities, or emotional intelligence, coaches provide targeted guidance and resources to support clients in acquiring new skills and refining existing ones.

3. **Support and Guidance**: A coach provides unwavering support and guidance to individuals. They create a safe and non-judgmental space where clients feel comfortable sharing their thoughts, feelings, and aspirations. By actively listening and understanding their clients' perspectives, coaches offer empathetic support and help them navigate through life's challenges.

4. **Goal Setting and Action Planning**: One of the primary functions of a coach is to assist individuals in setting meaningful and achievable goals. Through insightful questioning and exploration, coaches help clients clarify their objectives, whether they are related to personal growth, career advancement, health and wellness, or other areas of life. Once goals are established, coaches collaborate with clients to develop action plans outlining the steps needed to accomplish those goals.

5. **Obstacle Identification and Problem-Solving**: Life is

full of obstacles and setbacks, and coaches help individuals overcome these challenges with resilience and creativity. By exploring limiting beliefs, patterns of behaviour, and external barriers, coaches assist clients in identifying solutions and alternative perspectives. Through collaborative problem-solving, clients gain clarity and confidence in navigating obstacles on their path to success.

6. **Motivation and Inspiration:** Coaches serve as catalysts for motivation and inspiration, igniting client's passion and drive to pursue their dreams. Through encouragement, positive reinforcement, and strategic encouragement, coaches empower individuals to push past their comfort zones and strive for greatness. By fostering a supportive and motivational environment, coaches help clients stay focused and committed to their aspirations.

7. **Accountability and Progress Tracking:** Accountability is a cornerstone of the coaching relationship. Coaches hold clients accountable for their commitments and actions, helping them stay focused and disciplined. Regular checking and progress reviews allow coaches to assess client's development, celebrate milestones, and make adjustments to the coaching plan as needed.

8. **Holistic Well-being:** A holistic approach to coaching considers the interconnectedness of various aspects of life, including physical, emotional, mental, and spiritual well-

being. Coaches support clients in achieving balance and harmony across these dimensions, recognising that true fulfilment comes from nurturing all aspects of the self. By addressing both internal and external factors impacting well-being, coaches help individuals cultivate resilience, vitality, and a sense of wholeness.

9. **Reflection and Self-Discovery:** Coaching encourages introspection and self-discovery, allowing individuals to deepen their understanding of themselves and their aspirations. Through reflective exercises, journaling, and guided inquiry, coaches help clients gain clarity about their values, priorities, and purpose in life. This process of self-discovery fosters personal growth and empowers individuals to make intentional choices aligned with their authentic selves.

10. **Long-term Transformation:** Ultimately, the goal of coaching is to facilitate long-term transformation and sustainable change in individual's lives by providing ongoing support, guidance, and encouragement. Coaches empower clients to unleash their full potential and create meaningful, fulfilling lives. Through the coaching process, individuals develop the skills, mindset, and resilience needed to thrive in an ever-changing world.

In summary, the role of a coach in life encompasses a wide range of functions, all aimed at supporting individuals in realising their goals, overcoming challenges, and living their best lives. Through a collaborative and empowering partnership, coaches inspire clients to unlock their full potential and embrace their journey of growth and self-discovery.

Here, I (Maulik Choksi) take this opportunity of expressing my deep gratitude towards my Guru & Mentor shri shri Ravishankar, " Spiritual Leader" and Shri Sneh Desai, "Life & Business Coach" for being a source of inspiration, patience & kindness.

Please scan the Q.R. Code for further information:

"A good coach can change a game.
A great coach can change a life."

- John Wooden

Made in the USA
Monee, IL
05 November 2026

67808936R00144